FROGS AND TOADS OF THE SOUTHEAST

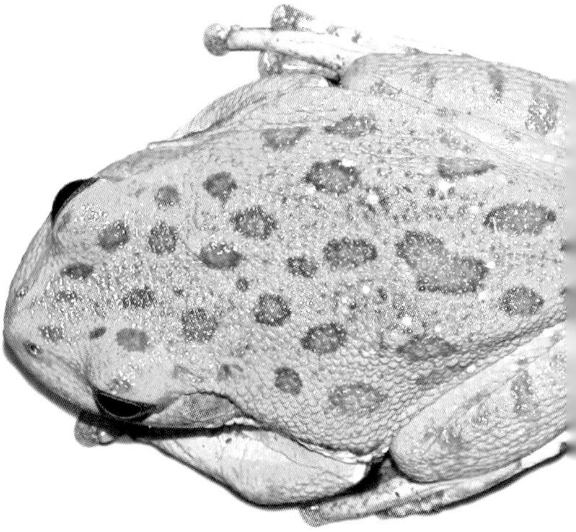

Frogs & Toads

OF THE SOUTHEAST

by Mike Dorcas and Whit Gibbons

The University of Georgia Press

Athens and London

Athens, Georgia 30602

Designed by Mindy Basinger Hill

Set in 10/15 Scala

Printed and bound by Four Colour Imports

The paper in this book meets the guidelines for
permanence and durability of the Committee on
Production Guidelines for Book Longevity of the
Council on Library Resources.

Printed in China

12 11 10 09 08 P 5 4 3 2 1

Library of Congress Cataloging-in-Publication Data

Dorcas, Michael E., 1963–

Frogs and toads of the southeast / by Mike Dorcas
and Whit Gibbons.

p. cm. — (A Wormsloe Foundation nature book)

Includes bibliographical references and index.

ISBN-13: 978-0-8203-2922-2 (pbk. : alk. paper)

ISBN-10: 0-8203-2922-3 (pbk. : alk. paper)

1. Frogs—Southern States. 2. Toads—Southern States.

I. Gibbons, J. Whitfield, 1939– II. Title.

QL668.E2D67 2008

597.8'90975—dc22 2008013582

British Library Cataloging-in-Publication Data available

Contents

FROGS AND TOADS OF THE SOUTHEAST

A squirrel treefrog

A young girl's excitement at capturing a toad

All about Frogs and Toads

WHY FROGS AND TOADS?

Frogs and toads are fascinating animals. Can you remember catching a toad in your neighborhood when you were a child? Or mucking around in a pond looking for tadpoles? As we grow up and our lives become filled with deadlines, mortgages, and other "grown-up" things, many of us lose the fascination we once had for the natural world and the animals that live in it. In the Southeast, frogs and toads are a vital and fascinating part of that natural world. Our goal in this book is to help readers recapture some of that fascination by reacquainting them with the world of frogs and toads. The book is a useful source of information about the biology of frogs and toads of the southeastern United States that we hope will increase everyone's appreciation for the uniqueness and importance of these animals as parts of our natural heritage.

Frogs and toads are easily distinguished from other animals. Anyone can identify an animal as a frog or toad, but few people are aware of the diverse and interesting lives these animals lead. Fortunately, the Southeast is the best place in the United States to become acquainted with frogs and toads because it is home to 42 species.

Frogs, toads, and other amphibians belong to an ancient group that was

A fossil frog (genus *Palaeobatrachus*) from the Late Oligocene (more than 20 million years ago) of the Czech Republic

present around 200 million years ago—well before dinosaurs roamed the earth. They were the first vertebrates to leave the water and live on land, thus paving the way for reptiles, birds, and mammals, including humans. They were the first vertebrates to have legs, lungs, and voices.

Frogs and toads are an amazingly successful group; more than 5,000 species are known, and many remain to be discovered. Unfortunately, some may not survive long enough to be described and named. It is a sad fact that many species of frogs and toads have recently gone extinct or are likely to do so shortly. More often than not, the disappearance of frogs and toads is either directly or indirectly the result of human activities. Many scientists think that the problems frogs and toads currently face are a harbinger of things to come for all of us. It is up to humans to save the frogs and toads and the planet on which we all live. The first steps in this task are education, learning to appreciate frogs and toads as part of our natural environment, and restoring the fascination we once had as kids.

DEFINING THE SOUTHEAST

The Southeast can be defined in many ways, including culturally, politically, physiographically, and biologically. Each type of definition uses different criteria for determining the boundaries of the region. Our book focuses on a particular taxonomic group—the frogs and toads—that is best studied and discussed in terms of physiographic regions and topography, climatic regime, and types of habitats. But because discrete margins are more practical for describing distribution across the landscape, we have based our geographic assignments on well-known political units—the states. We have defined the Southeast as comprising Alabama, Florida, Georgia, Louisiana, Mississippi, North Carolina, South Carolina, Tennessee, and Virginia. The Southeast so defined has an identifiable assemblage of frogs and toads that are clearly southeastern. Moving the boundaries farther north or west would include an array of species not typically associated with southeastern habitats.

CHARACTERISTICS OF FROGS AND TOADS

Frogs and toads are members of the class Amphibia. All amphibians are vertebrates—that is, they have a backbone—but not all amphibians are frogs or toads. The Amphibia also includes the salamanders (order Caudata) and the caecilians (order Gymnophiona), a group of legless, wormlike amphibians found primarily in the tropics.

The word *amphibian* literally means "both lives" and refers to the fact that most amphibians have an aquatic larval form (tadpole) that transforms into an adult, usually terrestrial, form. Although the ancestors of today's amphibians were the first vertebrates to move out of the water and invade the land, most amphibians remain tied to the water, requiring it for egg-laying and development of their larvae. Further, because of their permeable skin, many adult amphibians are susceptible to drying out and cannot survive in dry conditions.

Scientists have described more than 6,000 species of amphibians, and more species are discovered each year, especially in the tropics. More than 5,000 of the described amphibians are frogs and toads, a group collectively known as anurans (order Anura). The term *anura* literally means "no tail," and that is the key character distinguishing most

Frogs and toads are members of the class Amphibia. Other amphibians include the large, eel-like salamanders known as amphiumas (top), the spotted salamander (middle), and wormlike, tropical caecilians (bottom), which spend most of their time belowground.

Most frogs have hind limbs made for jumping.

Although often one of the first species of frogs to become active in winter, wood frogs sometimes get caught in below-freezing conditions and die.

adult anurans—the "tail-less" amphibians—from other amphibian groups. Most anurans have relatively long hind legs designed for jumping or hopping, another character that distinguishes them from other amphibians. Many anurans also have clawless, webbed toes on their hind feet.

ACTIVITY AND ENERGY

Like all amphibians, frogs and toads are cold-blooded, or "ectothermic"; that is, their body temperature is determined primarily by environmental conditions, not by internally generated body heat. Because they do not have to use energy to maintain a constant body temperature, frogs and toads can use much of their energy for growth and reproduction. They also have to eat less frequently than warm-blooded ("endothermic") birds or mammals and can thus survive longer periods without food.

Because frogs and toads are cold-blooded, however, they are active only when temperatures are warm enough to allow their bodies to function properly. During cold weather, many species hibernate in burrows or in the mud and debris at the bottom of ponds. However, several species of southeastern frogs, such as spring peepers, are actually more active during the winter and early spring than in the summer. Winter-active anurans usually spend the dry, hot summer months in an inactive state underground or under bark or leaves.

Treefrogs have pads on the ends of their toes that act like suction cups and allow them to climb.

Toads generally have warty skin, like this American toad. (left)
Frogs generally have smooth, wet skin, like this bullfrog. (right)

Most anurans, such as this barking treefrog, can't be categorized as either a frog or a toad.

When threatened, leopard frogs sometimes raise their forelimbs to their eyes, presumably for protection.

Many species of toads excrete a strong toxin from their skin glands. (above left) Pictured is a cane toad releasing poison from its parotoid gland.

Many frogs, such as this gray treefrog (above right), are well camouflaged.

DEFENSE MECHANISMS

Many frogs are food for other animals, and some animals actually eat nothing but frogs or toads. In many ecosystems, the combined biomass of anurans exceeds that of all mammals and birds combined. When they occur in large numbers, as they frequently do, frogs and toads can be extremely important components of the diet of many animals. Frogs and toads are not willing prey, however; they have developed unique tactics to avoid becoming a meal.

Many frogs and toads evade predators by leaping, and some can leap many times their own body length. Tadpoles of some species form large schools that, like herds of gazelles in Africa, reduce the threat of predation to any single individual. Most frogs remain hidden much of the time and thus avoid exposing themselves to potential predators. Even when

Frogs provide food for a variety of other animals, including snakes and insects. A garter snake (right) is making a meal of a green treefrog; a praying mantis (below) feeds on a young tree frog.

they are active, many species are well camouflaged and remain essentially invisible to a bird or other foraging predator.

The skin of many frogs and toads exudes toxic chemicals that can cause extreme irritation if they get in a predator's eyes or mouth. Toads have especially large skin glands. When grabbed, a toad can inflate its body and release a poison called bufotoxin from its parotoid glands. Bufotoxin contains a variety of compounds that can cause major physiological problems for natural predators as well as dogs, cats, and people. Some snakes that feed almost exclusively on toads have special enzymes that counteract the effects of the poisons.

REPRODUCTION

Nearly all southeastern frogs and toads go through three distinct life stages: egg, tadpole, and adult. Only the greenhouse frog and the coqui are exceptions; both complete the tadpole stage within the egg. All frogs reproduce sexually; that is, all require eggs contributed by a female and sperm from a male to produce offspring.

Some species of frogs, such as the southern leopard frog (left), have enlarged thumbs that help the male to hold on to the female during amplexus.

The desire to mate can be very strong in some species (right). Here a toad attempts amplexus with the photographer's finger.

Ornate chorus frogs in amplexus

Did you know?

Two species of Australian gastric brooding frogs care for their offspring by swallowing their eggs, which hatch and develop into little frogs within their mother's stomach. When fully developed, they crawl out of her mouth. Unfortunately, it appears that both species of gastric brooding frogs became extinct in the 1980s.

Reproductive activity begins with an advertisement call. Male frogs often call with many other males in large choruses that are audible at long distances. The calls attract females. In many species, females select males on the basis of particular call characteristics (pitch or call duration, for example). After she selects her mate, the female will usually approach the male closely, indicating that she is ready to mate. The male will then grasp the female under the armpits with his forelimbs; this mating grasp is known as *amplexus*. In eastern spadefoot toads, the significantly smaller male grasps the female around the waist instead of the armpits. Male frogs of many southeastern species have enlarged thumbs that help them to hold on tightly. Male frogs are sometimes so enthusiastic about mating that they attempt amplexus with other males or anything else that moves, even a person's ankle. Certain males of some species, called *satellite males*, may

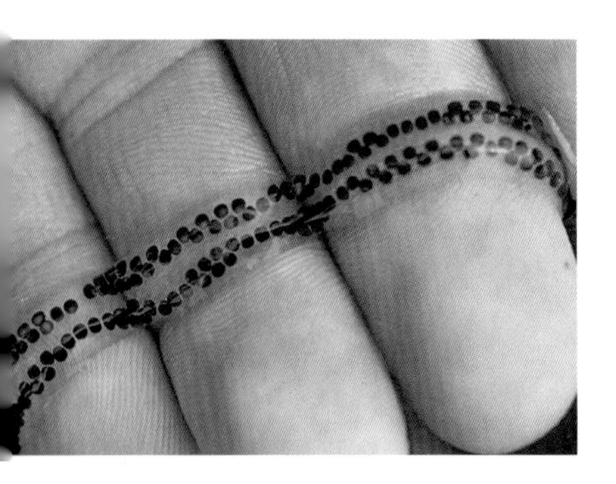

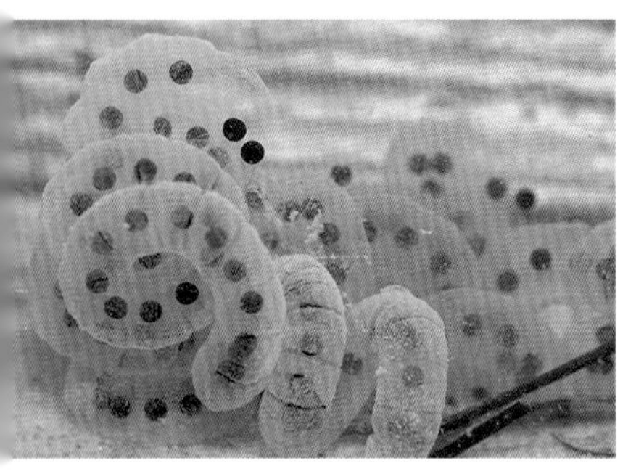

Leopard frog egg mass (top). Egg mass of eastern narrowmouth toad (top middle). Fowler's toad eggs (bottom middle). American toad eggs (bottom).

not call at all, but instead position themselves next to a calling male, then attempt to intercept and mate with females that are attracted to the calling male. This reproductive strategy allows them to obtain mates without the high energetic costs and risks associated with calling.

Egg-laying

Frog and toad eggs are almost always fertilized externally. After amplexus has begun, the female essentially carries the male to the water, where she lays eggs. As the eggs leave the female's cloaca, the male releases sperm that fertilizes them. The fertilized eggs absorb water, and the jellylike substance around them expands and helps to protect them from predators and desiccation. Toads lay eggs in long strings that are often draped across aquatic vegetation or lie on the bottom. Other southeastern frogs lay eggs in clumps or clusters of varying sizes; usually these are attached to emergent vegetation, but sometimes they lie on the bottom. When the tadpole developing within the egg is ready to hatch, the jelly-like egg capsule will disintegrate and the tadpole will swim free. Most frog and toad eggs hatch within a few days, although the eggs of some winter-breeding frogs, such as wood frogs, may take up to a month to hatch. The greenhouse frog and the coqui are once again exceptions. Both species have internal fertilization and lay their eggs in moist habitats on land. Larval development occurs within the egg, from which the fully developed froglets emerge.

In some species, like wood frogs, many females lay their eggs within a small area.

Tadpole and Metamorphosis

Tadpoles of most species grow rather quickly and in many cases undergo metamorphosis within 2–3 months. Tadpoles of species that breed in ephemeral wetlands tend to metamorphose earlier if the wetland is shrinking and about to dry up. If plenty of water remains in the wetland, the tadpole stage may be prolonged, allowing the tadpoles to grow larger before they metamorphose and move onto land. The most obvious signs of metamorphosis are the emergence of hind limbs at the point where the tail joins the body. Forelimbs emerge shortly after that. Once the limbs become more developed, the tadpole begins to undergo other dramatic changes. In many species, the entire digestive system changes from one designed for a herbivore feeding on algae to the digestive system of a carnivore. The gills begin to be resorbed as the metamorphosing tadpole develops lungs, and finally the long tail is resorbed. The tadpole usually leaves the water and ventures onto land while it still has a substantial tail. Metamorphosis is a hazardous time in the life cycle of frogs and toads. Not only are the froglets small and unable to move very effectively either in water or on land, their bodies are undergoing drastic anatomical and physiological changes as well. The combination of these factors makes them highly susceptible to predation.

American toad tadpole (top). A transforming river frog tadpole (top middle). A transforming green treefrog (bottom middle). Newly transformed green treefrogs (bottom).

Pine woods treefrog calling from a bush

Did you know?

Bats are major predators on some species of tropical frogs. The bats home in on the sounds made by male frogs when they are calling during the breeding season.

VOCALIZATIONS

Nearly all adult frogs and toads communicate with sound. Although the primary function of the male's advertisement call is to lure females, it can also warn encroaching males that they are within another frog's territory. In addition to the advertisement call, frogs produce several other sounds that deliver messages, including distress calls, release calls, and rain calls.

Recordings of a variety of calls from species featured in this book can be found at http://www.SEfrogs.org.

Sound Production

Sound production is closely linked to the frog's respiratory system. To produce sound, the frog forces air out of its lungs and into its throat, or buccal cavity, using strong contractions of the muscles surrounding the abdomen. Consequently, male frogs generally have more robust abdominal muscles than females do. As the air moves across the vocal cords, they vibrate, producing the call characteristic of the species. Many frogs have vocal sacs that provide resonating chambers to increase the volume of the call. Vocal sacs are usually either a single sac below the chin or paired sacs located on the sides of the head. Most frogs vocalize while perched on vegetation, sitting in shallow water, or floating on the surface. Some frogs that produce low-frequency sounds, such as the long, drawn-out snore of gopher frogs, may call primarily underwater, where low-frequency sounds carry well. Calling underwater likely also reduces the possibility of a predator locating a calling frog by cueing in on its call.

Advertisement Calls

Advertisement calls are generally species specific, and the distinctiveness of each species' call helps to prevent hybridization between species. Some species, such as cricket frogs and ornate chorus frogs, have short calls that resemble clicks or peeps; other species, such as American toads, have calls that may last more than half a minute. Some calls consist of a single note; others are a series of rapidly repeated notes, or trills. A frog's size does not necessarily indicate the volume of its call. Some very small frogs, such as spring peepers, can produce incredible volume, and some larger frogs, such as pickerel frogs, have relatively subdued calls. Females tend to listen for particular frequencies or call durations when selecting mates, and male frogs can increase their attractiveness by subtly varying the characteristics of their call.

Carpenter frogs have paired vocal sacs on either side of the head.

A satellite male green treefrog in amplexus with a female. The male intercepted the female as she was attracted to the calling male.

Calling Activity

The time of year or day and the weather conditions affect calling activity (see the table of calling months on page 219). In general, relatively warm, wet nights are the best time to hear frog choruses. Some species, such as chorus frogs and wood frogs, call during the winter or early spring when the weather is still fairly cold. In fact, wood frogs frequently call from the edges of ice-covered wetlands. Other species wait until later in the spring or summer to vocalize and reproduce. Many winter-calling species call during the daytime because nighttime temperatures are too low to allow calling. This "cold weather" activity allows them to breed in wetlands that may hold water only for short periods. Winter activity also reduces the likelihood of dehydration and predation by animals such as snakes that are active only during warmer periods.

Spring peepers sometimes call during the daytime.

Some species of frogs and toads may call at any time of the year if the weather conditions are right. For example, spadefoot toads are sporadic breeders that may emerge from their underground burrows anytime conditions are favorable, such as during and after prolonged thunderstorms. Males come to the surface and begin calling loudly from the small puddles and pools formed during the rain. Because the

Squirrel treefrog calling from a wetland in South Carolina

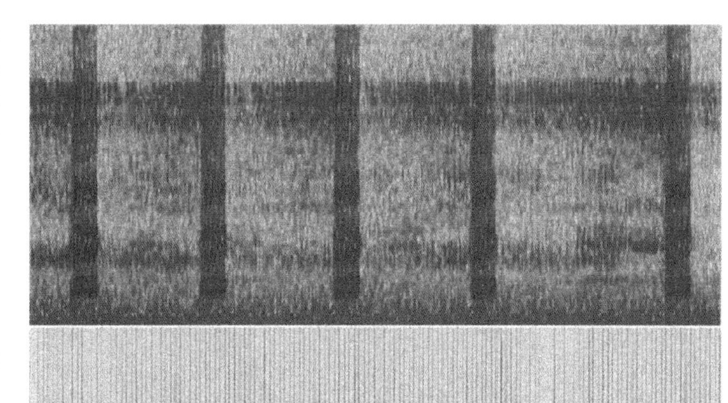

Sonograms of a Cope's gray treefrog (top) and a southern toad (bottom). Note how the short trills of the treefrog contrast with the long, soft call of the toad.

puddles and pools can dry within a few days, spadefoot toad eggs and tadpoles develop quickly so that the toadlets can emerge before the wetland dries.

Other Types of Calls

In addition to advertisement calls, frogs and toads produce other types of calls for various purposes. Some species have distinctive territorial calls to dissuade intruding males from entering the calling male's territory. Bullfrogs emit a loud and abrupt *phoot* when another male ventures too close, and green treefrogs produce a rapidly repeated call when another male green treefrog is nearby that is very different from their usual *quenk*. Many "true frogs" such as green frogs, leopard frogs, and bullfrogs give a loud squawk as they leap into the water from the bank of a pond. Some frogs produce distress calls when captured by a would-be predator. Bullfrogs and leopard frogs, for example, produce a loud, high-pitched, open-mouthed scream as they are being swallowed alive by a snake. Scientists think the purpose of the scream is to attract other potential predators that might distract the snake and allow the bullfrog to escape. Squirrel treefrogs produce a rain call distinct from their usual advertisement call in response to an approaching rainstorm, presumably signaling females of a potential breeding opportunity.

Calling Energetics

Calling is among the most energetically demanding activities of frogs. For many species, vocalizing actually requires more energy than repeated leaping for the same amount of time. Male frogs have special physiological modifications in their abdominal muscles that support sustained energetic expenditure. The muscle cells have many mitochondria, the cell structures that produce energy from sugar, and extensive blood vessels to carry plenty of oxygen to the muscles.

A red-eyed treefrog from Costa Rica

Did you know?

*Skin toxins from the golden poison dart frog (*Phyllobates terribilis*) are so potent that merely touching the tip of your tongue to the back of the frog can be fatal.*

DIVERSITY AROUND THE WORLD

A streamlined, smooth-skinned green treefrog with its racing stripe and long, graceful legs is easily distinguished from a dumpy, warty American toad with its large poison glands and short legs. But both have two features in common that distinguish them, and all other frogs and toads, from other major groups of animals: their hind legs are appreciably longer than their front legs, and adults lack a tail. Other animals may possess one of these traits, but frogs and toads always have both. Nonetheless, the anurans offer an array of diverse and bizarre behaviors, fascinating ecologies, and captivating appearances. If for no other reason, they are remarkable for their incredible diversity of reproductive strategies.

Anurans live on virtually every large island and continent in the world except Antarctica. Comparing the numbers found in tropical and temperate areas is a good way to gain perspective on their distribution. For example, when the half-dozen largest countries in the world are compared, Can-

Some South American Indians have used secretions from poison dart frogs to tip the ends of blowgun darts.

ada and Russia each have fewer than three dozen species of frogs and toads, the United States has about 100, Australia has more than 200, and China has almost 300. But those numbers seem paltry when compared with Brazil's more than 700 kinds of frogs and toads, and probably many yet-to-be-discovered species as well. On a smaller scale, France has only 27 species of frogs and toads, whereas Ecuador, at half the size, has 430. Although the greatest species diversity is in the tropics, the wood frog can live north of the Arctic Circle, surviving winter hibernation by converting its body fluids into a form of antifreeze, and the Himalaya frog (*Nanorana parkeri*) is found at elevations above 16,000 feet.

Frogs may not vary much in body shape, but they display a seemingly endless spectrum of lively colors. The Oriental fire-bellied toad (*Bombina bombina*), green above and red below, is startling, and the rare blue color phase of the Australian green treefrog (*Litoria caerulea*) is intriguing, but the bright warning coloration of the blue poison dart frog (*Dendrobates azureus*) of Suriname would make any bird think twice before trying to eat it. Other poison dart frogs of the tropical family Dendrobatidae feature spectacular combinations of brilliant reds, oranges, yellows, greens, blacks, and whites.

Perhaps the single most remarkable feature of frogs is their wide range of reproductive tactics. At one extreme is the western Nimba toad (*Nimbaphrynoides occidentalis*) of Africa. Females bear live young without laying eggs, nourishing the embryos within their body until they become fully developed toadlets. Parental care, often by the male, is common in several families of frogs and toads. The female Darwin's frog (*Rhinoderma darwinii*) of South America, for example, lays about 40 eggs on land. The male fertilizes them and remains with them, and after several days, he gulps them into his mouth and puts them into his vocal sacs. The eggs may take as long as 50 days to develop into froglets, at which time he lets them hop out of his mouth onto the ground. The male of the closely related Chile Darwin's frog (*Rhinoderma rufum*) carries the eggs in his vocal sacs for

Anurans offer an array of captivating appearances. Pictured here are a horned frog from South America (*Ceratophrys cranwelli*, top) and a Java flying frog (*Rhacophorus reinwardtii*, bottom). Species of flying frogs use the extensive webbing between their toes to glide from tree to tree.

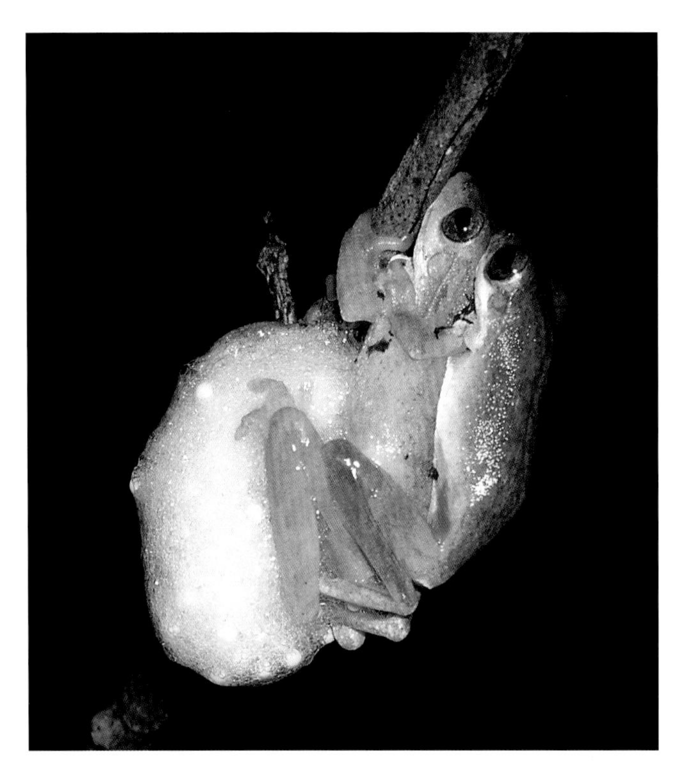

Some African frogs construct a nest of foam on a branch overhanging water in which they lay their eggs. The outer layer of foam hardens to protect the developing eggs from drying out.

only a couple of weeks, then takes them to water and lets the tadpoles drop out of his mouth; the tadpoles finish their development in the water. The male of the completely aquatic Surinam toad (*Pipa pipa*) acts as a midwife, placing the eggs the female releases into the water onto her back, where her skin grows over them. The mother toad goes about her business for as long as 4 months before the young toadlets break through the skin on her back and emerge into the water.

Did you know?

The eggs of the strange-looking Surinam toad grow under the skin on the mother's back. The male helps to place them there, and the skin grows over the eggs to protect them. The eggs and tadpoles develop under the female's skin, and the fully developed toadlets break out.

Wallace's flying frog (*Rhacophorus nigropalmatus*) of Malaysia is remarkable not only for its huge webbed feet, which allow it to glide from tree to tree like a flying squirrel, but also because it is one of the many species of foam-nesting frogs. When the female is ready to nest, she climbs into a tree overhanging water, releases fluid from her body, and then kicks her legs to whip up a foamy froth. She lays her eggs in the foam and the male fertilizes them. The embryos within the eggs develop into tadpoles at about the same time that the nest begins to disintegrate, so that the tadpoles fall into the pool of water below and complete their development.

The tropical frogs are indeed amazing, but many frogs and toads of the Southeast are special in their own right. Several species of chorus frogs, including the aptly named ornate chorus frog, call and breed in southeastern wetlands in the winter. Pig frogs engage in hand-to-hand combat when one male enters the territory of another. Spadefoot toads may spend more than a year underground waiting for the right conditions to breed. As adults, little grass frogs are only the size of a person's fingernail and produce a call so high-pitched that many people cannot even hear it. To be sure, the Southeast, which harbors nearly half of the frog and toad species found in the United States, has its share of anuran intrigue.

INTRODUCED SPECIES

Individuals of many species move away from their place of birth to inhabit new areas. One consequence of this dispersal is the colonization of new regions. Natural dispersal from one continent or island to another has created many of the patterns of animals' geographic distribution. In today's world, however, people are responsible for transporting many species to new areas. Humans have introduced animals and plants to regions outside their natural geographic range for centuries. Some of the species introduced into the United States were brought intentionally, such as honeybees from Europe; others, such as fire ants from South America, were brought unintentionally. No matter where they are introduced to or from, most exotic species do not survive, or their presence is benign or even goes unnoticed. But species that have no natural enemies in the new region can become invasive, as fire ants have, causing problems for humans, native wildlife, or both.

Invasive Frogs

Ironically, among the species of invasive animals that have had noticeable negative impacts on our native frogs are other frogs, namely the Cuban treefrog and the cane toad (see species accounts). Whether the Cuban treefrog was transported unwittingly from Cuba to Key West, Florida, by boat prior to the 1930s or whether it arrived naturally will probably never be known, but its rapid progress northward up the Florida peninsula has certainly been aided by humans. Cuban treefrogs are notorious for eating other frogs, including a variety of native species.

Cuban treefrogs can be found throughout much of Florida.

The cane toad was first introduced into southern Florida in the 1930s with forethought, to control pests in sugarcane fields. Later, probably in the 1950s, a shipment of these toads believed to be the forerunners of those that ultimately expanded the geographic range of the cane toad in Florida escaped near the Miami airport. Cane toads eat native frogs but create an additional problem because house pets or wildlife that bite them can become ill or even die from the toxins produced by the toad's parotoid glands. The cane toad was intentionally introduced into Australia in 1935 to control pests in the cane fields and within a few decades became a major threat to native wildlife. Extensive efforts to eradicate the toad from Australia continue.

Two other frog species that are now established in southern Florida are the greenhouse frog and Puerto Rican coqui (see species accounts). Neither is known to cause environmental problems for native frogs or other wildlife; however, some residents of Hawaii, where the latter species is also introduced, consider the loud advertisement call of the Puerto Rican coqui to be "noise pollution." The Rio Grande chirping frog (*Syrrhophus cystignathoides*) has been introduced into Shreveport and Baton Rouge, Louisiana, and may have established reproducing populations in both cities. Like the greenhouse frog and coqui, it is not known to cause any environmental problems.

Some Australians have found ways to entertain themselves with introduced cane toads.

At least one southeastern frog, the bullfrog, has become an invasive species in other areas of the United States and elsewhere. Bullfrogs have fewer natural predators in the western states than they have in the East, and some amphibian biologists hold them responsible for reducing the numbers of other frogs, including the rare Chiricahua leopard frog (*Rana chiricahuensis*) and lowland leopard frog (*Rana yavapaiensis*). Not all herpetologists blame the bullfrog for the decline of the two leopard frogs, but bullfrogs certainly appear to have become the dominant frog species in areas where they did not occur before.

African clawed frogs (*Xenopus laevis*), originally brought to the United States in the mid-1900s for use in human pregnancy tests, have been in-

Cane toads have caused many problems for native fauna in Australia, where they were introduced decades ago. This introduced species can be extremely abundant in south Florida.

Bullfrogs are natives of the Southeast that have caused major problems in the western United States.

tentionally or accidentally released into the wild in several southeastern states, including Florida, North Carolina, and Virginia. They have established populations only in California and Arizona, however, where some conservation biologists consider them a potential threat to environmentally sensitive fish and amphibians.

Invasive Fish

Among the most severe threats to our native frogs are trout, bass, and other predaceous game fish that have been introduced into lakes and ponds where amphibians breed. These fish eat the eggs and tadpoles of many species of frogs and toads, contributing to overall population declines. Some larger predatory fish also eat adult frogs. Placing game fish in formerly fish-free lakes can lead to the complete elimination of frog populations. Introduced sunfish and bass have caused substantial reductions of frog and toad populations in the Southeast.

Other Invasive Species

Numerous other nonnative species directly or indirectly affect frogs and toads. Feral cats and house cats allowed to roam outdoors probably kill some frogs, but whether they cause noticeable changes at the population level is unknown. Also undetermined is the effect on anurans of introduced exotic plants that can change the plant composition and structure of wetland habitats.

What Can Be Done about Invasive Exotics?

What can we do to halt the introduction of other exotic animals and plants or reduce their environmental impacts? The tropical climate of southern Florida is likely to be favorable to other exotic species that are intentionally or accidentally released, and the Florida peninsula could serve as a corridor to the remainder of the Southeast. We offer two suggestions.

Did you know?

Bullfrogs have been introduced into many areas of the western United States, where they have become an invasive species that eats native wildlife, including many other frogs.

The Rio Grande chirping frog is native to southern Texas but has been introduced into Shreveport, Louisiana.

The implementation of laws or regulations that impose strict controls on the importation of exotic wildlife species is one step, particularly in Florida and warm-weather ports in the Gulf of Mexico. Entry into this country of exotic species that could cause problems for native frogs and toads if they escaped or were intentionally released must be given careful scrutiny. The time has probably passed for controlling the spread of the Cuban treefrog and cane toad as they march northward through Florida and into the rest of the Southeast. We must wait to see if natural controls in the forms of parasites, predators, or dramatically cold weather become factors curtailing their spread and reducing their numbers. The focus must be on preventing the introduction of even more anuran species as potentially devastating to other frogs as the Cuban treefrog and cane toad have been.

Second, a reasoned approach is necessary when stocking predaceous game fish in amphibian breeding sites. Most state wildlife and natural resources workers understand the problems of focusing only on game species without regard for the thousands of other species that make up natural ecosystems, so that threat to amphibians may soon be a phenomenon of the past.

IDENTIFYING FROGS AND TOADS

One of the first steps in developing an appreciation for the natural world is being able to identify the animals and plants you encounter, including frogs and toads. Once you know the species of the frog or toad you have just seen or heard, you can find out how large it can get, how many eggs the females lay, and what the tadpoles look like. Most of the 40 or so species of frogs and toads native to the Southeast can be identified by using a few key characters. Learning such characteristics is easy and can greatly add to your enjoyment of nature in general and frogs and toads in particular. Knowing where a frog came from can be especially helpful, too (e.g., a brown frog from North Carolina or Virginia with four yellow stripes down the body is without question a carpenter frog, and a stocky frog with relatively short legs and distinct dorsolateral ridges from western Louisiana is a crawfish frog rather than one of the gopher frogs).

Among the visible characters commonly used to differentiate among certain species of frogs and toads are enlarged pads on the toes (the treefrogs and some chorus frogs), extensive webbing between the toes of the hind feet (the true frogs), and conspicuous enlarged warty structures (parotoid glands) on the top of the head (the true toads). Like birds, each southeastern frog and toad species has a characteristic, distinctive repertoire of calls. Often the call alone is sufficient to confirm the presence of a species.

Even the tadpoles of most southeastern species have characters that set them apart from other species. A person familiar with tadpoles will immediately recognize a solid black 6-inch-long tadpole with bright red eyes from South Carolina or Georgia as a river frog. But although body color can be helpful in identifying some tadpoles, many tadpoles of different species resemble one another to the casual observer. One certain way to identify tadpoles is to examine the mouthparts, which are distinctive for each species. But features such as the number of labial tooth rows or whether the oral disc is present are subtle characters used primarily by professional herpetologists (with a microscope) and are not practical for identifying a live tadpole caught in a wetland.

The following guide to the species of frogs and toads of the Southeast provides a combination of characters—some quite obvious and others requiring closer examination—that can be used to distinguish each species from all or most of the others. For tadpoles, we provide general descriptions of the typical color, size, and shape. The identification of adults and

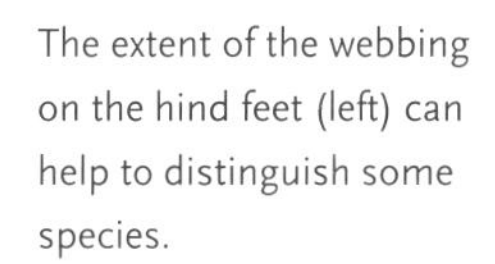

The extent of the webbing on the hind feet (left) can help to distinguish some species.

The mouthparts of tadpoles (right) are among the main characters used to distinguish among species.

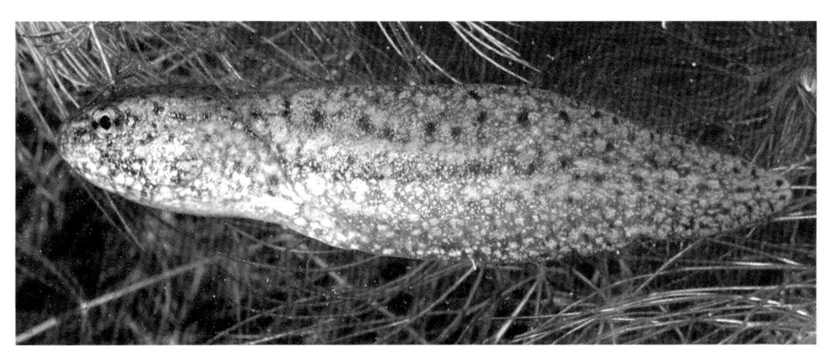

The size and shape of the tail fin can help to identify tadpoles.

The yellow groin and thighs distinguish the two species of gray treefrog from other treefrogs.

tadpoles of some species will require careful examination by an expert in herpetology. And even professional herpetologists will sometimes have difficulty confirming whether the small olive brown frog in their hand is an immature pine woods treefrog or an adult squirrel treefrog. But most southeastern frogs and toads can be correctly identified with minimal effort by using the information provided in this field guide.

Colors and Patterns

Southeastern frogs and toads are typically less colorful than many birds, snakes, and butterflies. Although some frogs and toads display reds and yellows, the adults and tadpoles of most species wear the colors of camouflage—brown, green, and black. Spotting patterns, stripes, dark face masks, and splashes of yellow, orange, or green concealed beneath the hind legs are all useful characters. General body color is not always a clue to identity, though, because individuals of many species change from light colored to dark in response to changes in temperature, time of day, and activity level. Some of the treefrogs are green under one set of environmental conditions and brown or gray under another. The cricket frogs and ornate chorus frogs may display a variety of color patterns within a single population, and some species of true toads vary from red to brown to black in the same area.

Did you know?

Frogs that produce poisonous skin secretions are often brightly colored as a warning to potential predators.

Size

Southeastern frogs come in a broad range of sizes. A full-grown little grass frog can sit comfortably on a dime, while the massive bullfrogs and pig frogs are big enough to cover most of a dinner plate. Toads range in size from narrowmouth toads and oak toads that are less than 2 inches long to the giant cane toads of Florida that can reach 8 inches in length. Herpetologists use the snout-to-vent length to measure the body length of frogs and toads. This simple and straightforward measurement is the straight-line distance from the tip of the snout to the cloacal opening. Tadpole size is generally measured from the nose to the tip of the tail (as adult anurans have no tail, tail length is not an issue for them). Knowing the size of an individual may help to rule out other species that never get as big. Adult females of many species grow appreciably larger than males. In only one southeastern species, the bullfrog, does the male grow larger.

Key Traits

A few characteristics are especially useful in identifying frogs and toads.

PAROTOID GLANDS AND CRANIAL CRESTS Cranial crests and parotoid glands are particularly useful in differentiating true toads (genus *Bufo*) from other species and from each other. The cranial crests—solid ridges alongside and behind the eyes—are typically in front of the elevated and very apparent pair of oval parotoid glands. The shape of the cranial crests and their configuration relative to the eyes and parotoid glands—which also vary in shape, size, and position—are distinctive for many species of true toads. Other frog and toad species in the Southeast have no cranial crests, and parotoid glands, if present, are inconspicuous.

The shape of the cranial crests helps to distinguish the various species of toads.

TOE PADS Treefrogs and some chorus frogs have enlarged toe pads on the front and back feet that form adhesive disks useful for climbing. The proportional size of the disk varies among species and in some cases can be used as a distinguishing character.

The toe pads of a treefrog distinguish it from most other southeastern species.

Most species of frogs have horizontal pupils (top), but spadefoot toads have vertical pupils (bottom).

BODY SHAPE Frogs are generally more streamlined than toads, which tend to be short and squat. Subtle differences in body shape, such as the teardrop shape of the narrowmouth toad or the slim-waisted appearance of a green treefrog, may also be useful.

BODY COLOR PATTERNS Some species can be identified by their body markings alone. The blotches on the back of a pickerel frog are noticeably squarish when compared with the more rounded blotches of a leopard frog. Light stripes on a dark background or dark stripes on a lighter background are signature markings for some species, such as the two lines on the thigh of a Florida cricket frog or the dark and light lines down the sides of a Coastal Plain toad.

SKIN The texture of the skin—smooth, rough, or granular—readily differentiates many species. The presence and location on the body of warts or small tubercles can be important for distinguishing some species.

ADVERTISEMENT CALLS Among the most important characteristics that distinguish frog or toad species is the advertisement call of the male. Species that appear to be identical, such as Fowler's toad and the southern toad, can be immediately distinguished by their unmistakably different calls. In fact, the two species of gray treefrogs that are identical in appearance can be differentiated in the field only by their calls. Recordings of calls can be found at http://www.SEfrogs.org.

OTHER DISTINCTIVE CHARACTERS Shape of the eye pupils, protuberances on the hind feet, and the presence of elevated ridges down the back are other traits that can be useful in differentiating among species. Certain behaviors are also characteristic of particular species or groups of species. For example, any frog might be found on the ground, but a southeastern frog clinging to the trunk of a tree will almost certainly be one of the treefrogs in the genus *Hyla*. A species that gives a loud chirp as it makes a long jump from shore into the water will most likely be one of the true frogs (genus *Rana*).

GEOGRAPHIC LOCATION The region of the Southeast where a frog or toad is found often provides an instant clue to its identity. Wood frogs, for example, are absent from Louisiana, Mississippi, and Florida; Strecker's chorus frogs are not found east of Louisiana; and the bog frog is restricted to Florida. Identification of a local species can often be narrowed considerably by eliminating those species that do not naturally occur in the region.

HABITAT The habitat where the animal was found can be another important clue to its identity. Most true toads are likely to be found on land, whereas pig frogs are usually in the water. The specific location within a habitat may also provide meaningful information. For example, a leopard frog is not likely to be in a bush or tree.

TIME OF YEAR The season can provide significant information because some species of frogs and toads have well-defined windows of breeding activity and are extremely difficult to find at other times of the year. In some areas of the Southeast, the winter-breeding chorus frogs are often the only frogs that might be present and calling from a wetland on winter nights.

TIME OF DAY Most species of frogs and toads are active at night, and daytime appearances are rare. Small frogs hopping around the edge of a wetland during the day would typically be cricket frogs, juvenile true toads, or spadefoot toads rather than treefrogs.

Dorsolateral ridges that extend the length of the body distinguish a green frog from a young bullfrog.

Ornate chorus frog
Pseudacris ornata

Spring peeper
Pseudacris crucifer

NAMING FROGS AND TOADS

Taxonomy is the scientific field of the classification and naming of organisms. Anuran taxonomists classify and name frogs and toads in a way that reflects the ancestral relationships among species. Thus, closely related species are placed together within a genus (plural = genera), and closely related genera are grouped together within a family. All animals have a two-part scientific name, which is always italicized. The first name is the genus name (e.g., *Pseudacris*), which is capitalized. When the lowercase *crucifer* is added after the genus name, we have the scientific name of the spring peeper, *Pseudacris crucifer*. The species name of the ornate chorus frog is *Pseudacris ornata*, and that of the upland chorus frog is *Pseudacris feriarum*. The placement of these three species in the same genus indicates that they are more closely related to each other than to other frogs that are in the same family but in other genera, such as *Hyla* and *Osteopilus*.

A Word about Frog and Toad Taxonomy

Taxonomy is not a perfect science, and taxonomists often disagree about the relative importance of different traits. Modern molecular genetics has helped herpetologists to resolve many taxonomic debates. Nonetheless, differing viewpoints about the lineage and ancestral relationships (phylogeny) of particular species or groups of species remain. Sometimes these disagreements result in the scientific name of a species being changed. In this book we have used the scientific names familiar to most herpetologists, although other names may have been proposed. This is not a perfect scheme, because some scientific names (such as that for the spring peeper, which was changed during the latter half of the twentieth century) are accepted immediately while other taxonomic revisions may not be fully accepted by the herpetological community for years. Two familiar southeastern species, the American toad and the bullfrog, are ideal examples of such a situation and how we have chosen to handle it.

In the early 2000s, amphibian biologists began taking advantage of new molecular biology techniques to reexamine the phylogenetic relationships of all frogs and toads, an important and worthwhile exercise for understanding genetic, behavioral, and other biological similarities and dissimilarities among taxonomic groups. One consequence of these studies was a reinterpretation of the relationships of the American toad and the bullfrog to other species within their own family and genus. Because of the strict rules of scientific nomenclature, which we need not dwell on here, the new names proposed were *Anaxyrus americanus* for the American toad and *Lithobates catesbeianus* for the bullfrog. Herpetologists have used the scientific names of the American toad (*Bufo americanus*) and the bullfrog (*Rana catesbeiana*) since the 1800s, though, and all the scientific literature concerned with these species refers to them by those scientific names. Changing the names is bound to cause a certain amount of confusion, which we hope to avoid here. In this book we continue to use the more familiar, traditional scientific names by referring to the American toad as *Bufo americanus* and to the bullfrog as *Rana catesbeiana*. The same is true for other southeastern species of frogs and toads whose scientific names have been recognized and accepted by amphibian biologists for decades or centuries. The species accounts discuss taxonomic changes and include names that have been proposed and that may become widely used in the future. We include a short explanatory note in the accounts for species whose classification was in dispute at the time of writing. But our primary goal is to be certain that the reader knows what frog or toad we are talking about, no matter what scientific name someone thinks the animal should be given.

Upland chorus frog
Pseudacris feriarum

A young pig frog in a southern swamp

Species Accounts

INTRODUCTION

The following species accounts are designed to familiarize the reader with every species of frog and toad native to the Southeast. The accounts are grouped into four categories based on the current classification of frogs and toads. The four categories are (1) members of the treefrog family, Hylidae; (2) the true frogs in the family Ranidae; (3) the true toads in the family Bufonidae; and (4) a catchall group that includes the spadefoot toads (Pelobatidae), the narrowmouth toads (Microhylidae), and the introduced species of the family Leptodactylidae.

Each species' geographic range in the Southeast is indicated on the map that accompanies its account. A smaller map shows the entire U.S. range for species that occur outside the Southeast. The range maps are based on a combination of current and historical records. The shaded range should be viewed as an approximation of the actual presence of a species, as almost no frog has a continuous distribution across all habitats within a region. As an example, the range map for southern leopard frogs indicates that they occur throughout the Southeast. Their actual distribution is patchy, however, because of their dependence on wetland habitats and the disappearance of local populations due to natural or human-based causes.

A calling little grass frog

BASIC FEATURES OF THE SPECIES ACCOUNTS

The species entries are arranged as follows (not all elements occur with every entry):

The oak toad characteristically has a well-defined light stripe down the middle of the back.

Quick identification guide

How do you identify an oak toad?

SKIN
Rough

LEGS
Relatively short

FEET AND TOES
Hind toes unwebbed

BODY PATTERN AND COLOR
Gray with brownish markings on either side of a whitish or yellowish stripe; belly whitish

DISTINCTIVE CHARACTERS
Small size; light-colored line down back

CALL
Repeated, high-pitched peeps sounding like a baby chicken

SIZE
max tadpole = 1.5"
typical adult = 1"

Oak Toad *Bufo quercicus*

DESCRIPTION Oak toads are the smallest true toads in the United States; the largest individuals are less than 2 inches long. The body color is usually light to dark gray, with an occasional brownish tint. Large, widely spaced brown or black blotches on the back sometimes form broad lines along the sides. A very distinct light line that is often white but may be yellowish extends down the center of the back. The gray legs have broad, dark bands. The cream-colored belly has a granular appearance. The parotoid glands are large and extend from either side of the back of the head down the neck, but they may be less obvious than those of other toads because the oak toad is so small.

The adult oak toad is only slightly more than 1 inch long.

WHAT DO THE TADPOLES LOOK LIKE? Like the adults, the tadpoles are small, usually less than 1 inch long, and may be gray, dark greenish brown, or almost black with a whitish belly. The mouth and snout turn downward. The tail fin is transparent, but the central, muscular area of the tail appears to have light and dark bands.

172 • *Oak Toad*

Large map showing where this species is found in the Southeast

Oak Toad
Bufo quercicus

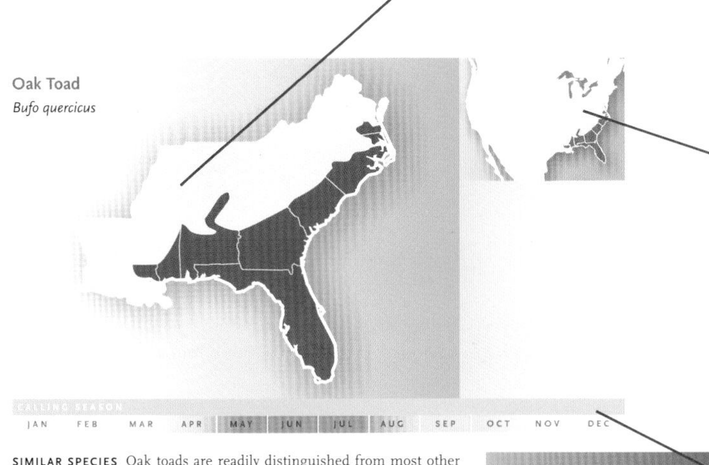

Small map with a shaded area showing where this species is found in the continental United States

CALLING SEASON

JAN FEB MAR APR MAY JUN JUL AUG SEP OCT NOV DEC

Greater intensity of shading indicates when a species is likely to call.

SIMILAR SPECIES Oak toads are readily distinguished from most other toads by their small size and the prominent light center stripe, which contrasts sharply with the gray body. Southern toads occasionally have a hint of a light line down the back, but they have cranial crests in front of their parotoid glands. Coastal Plain toads with a light stripe down the center also have a distinct light-colored stripe down each side with a broad, dark stripe below it.

DISTRIBUTION AND HABITAT Oak toads are found only in the Southeast, ranging throughout Florida and in parts of every other southeastern state except Tennessee. Their distribution is limited primarily to the Coastal Plain but extends north into central Alabama. Oak toads are typically found in association with pine woods or mixed oak and pine habitats, often with grassy areas. They breed in a variety of wetland situations including Carolina bays, freshwater marshes, cypress ponds, and flooded ditches and low areas.

BEHAVIOR AND ACTIVITY Like most other frogs and toads, oak toads are active at night, but they are also active—and most often observed—during the day, especially under moist conditions during the warmest months. During unfavorably dry or cold weather they retreat to underground burrows or beneath logs and other ground litter. They sometimes live in terrestrial habitats as much as 250 feet from the wetland breeding sites to which they migrate during rainy periods. They are inactive during cold weather from late fall to early spring.

Oak Toad • 173

Did you know?

Toads have warts, but handling a toad cannot give a person warts.

Interesting or anomalous facts about this species or about frogs in general

CRICKET FROGS, CHORUS FROGS, AND TREEFROGS

A male northern cricket frog calling to attract a mate

How do you identify a northern cricket frog?

SKIN
Warty

LEGS
Muscular and long

FEET AND TOES
Hind feet webbed

BODY PATTERN AND COLOR
Usually some combination of brown, gray, green, yellow, or black; hind legs with dark stripes

DISTINCTIVE CHARACTERS
Dark triangle between eyes; hind leg stripe with irregular margin

CALL
Rapid series of clicks

SIZE
max tadpole = 1.75"
typical adult = 1"

Northern Cricket Frog *Acris crepitans*

DESCRIPTION Northern cricket frogs are small and variably colored. The body can be a mixture of brown, green, or gray and may have overlying markings of yellow or black, but usually not red. Some individuals are a solid color. The belly is grayish or white. The head, body, and legs have small warts, giving the skin a granular appearance. The maximum body length of adult females, which are slightly larger than males, is less than 1.5 inches. The head of the northern cricket frog is relatively rounder and blunter than that of the southern cricket frog. A dark triangle whose apex points toward the rear is usually present between the eyes. A dark stripe runs along the back of the thigh. The webbing between the first and second toes is extensive and the legs are proportionately longer for their body length than the legs of any other North American frog, including the very similar southern cricket frog. Because southern and northern cricket frogs commonly interbreed, distinguishing individuals as one species or the other can be difficult in areas of geographic overlap.

The irregular dark stripes on the rear of the thighs of northern cricket frogs distinguish them from southern cricket frogs.

Northern cricket frogs are often found on land along the muddy shores of lakes, ponds, and rivers.

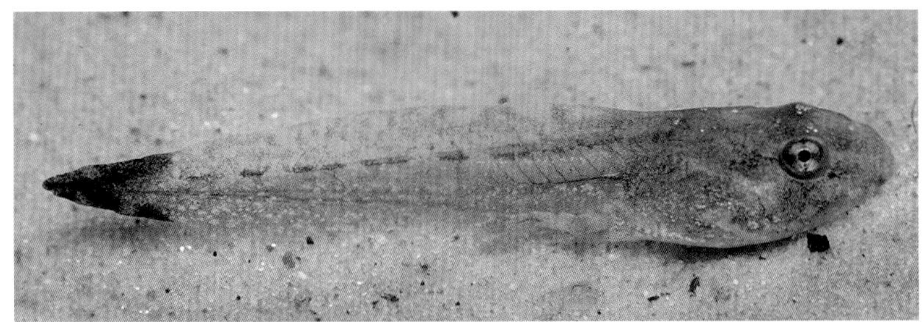

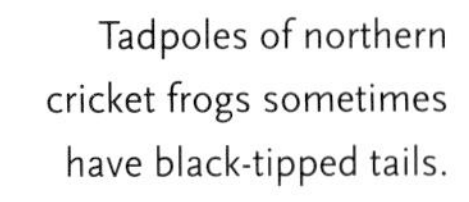

Tadpoles of northern cricket frogs sometimes have black-tipped tails.

WHAT DO THE TADPOLES LOOK LIKE? Northern cricket frog tadpoles cannot be distinguished from the tadpoles of southern cricket frogs on the basis of body color (light to dark greenish brown) or the long, translucent tail with dark spots or speckling. The tip of the tail can be solid black in either species. The largest northern cricket frog tadpoles can reach a total length of 1.75 inches.

SIMILAR SPECIES Its smaller size and more warty skin, the dark triangle on the head, and the single dark stripe on the rear of the hind leg separate the northern cricket frog from most other frogs. Northern cricket frogs also have more webbing on the hind feet and a broader, blunter head, and the leg stripe has an irregular border that is more jagged than that of the southern cricket frog.

DISTRIBUTION AND HABITAT The northern cricket frog is found in all or part of every southeastern state, and its overall geographic range covers most of the eastern United States and extends from southern Canada into

Northern Cricket Frog

Acris crepitans

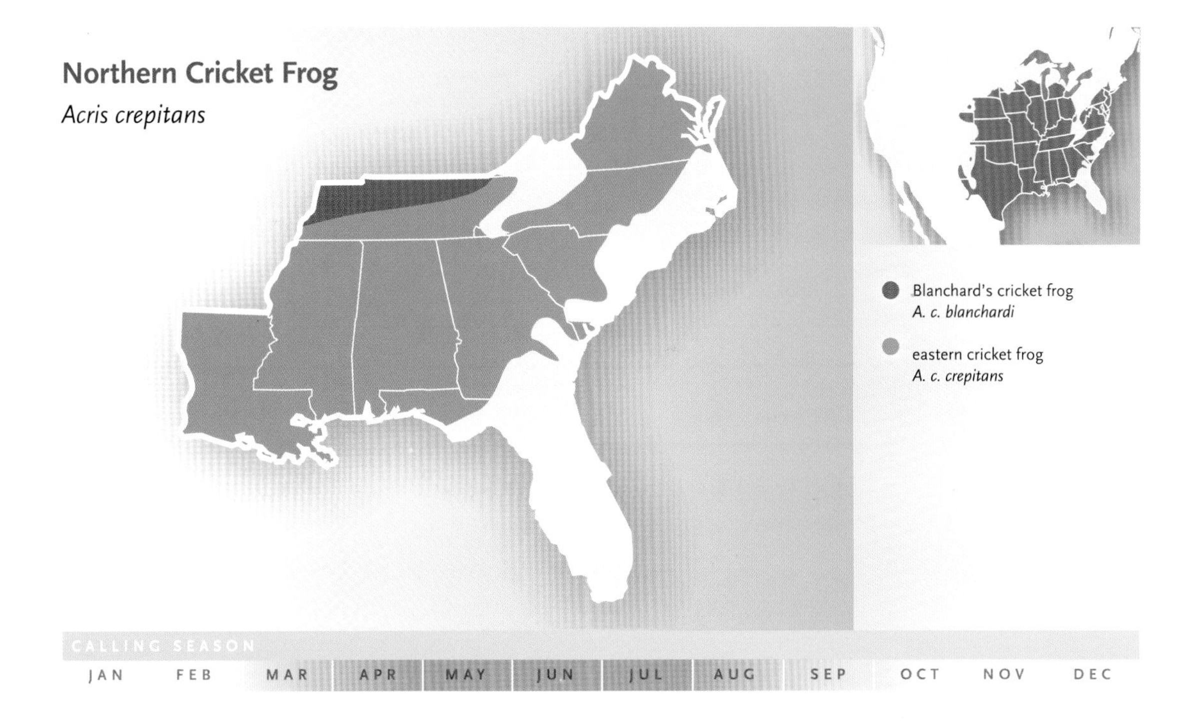

Mexico. The species is absent from most of Florida and much of the lower Coastal Plain of the Carolinas. Northern cricket frogs occupy areas in and around freshwater habitats, including swamps, wetlands, farm ponds, small streams, and bogs, and may be abundant around permanent water bodies as well as ephemeral wetlands. They are especially abundant in open habitats alongside wetlands but tend to be less common around large rivers and lakes that lack shallow, vegetated margins.

BEHAVIOR AND ACTIVITY In the Southeast, northern cricket frogs can be seen during the day or night in the vicinity of aquatic habitats, but they are more active in the daytime during cool periods. During dry or cold weather they retreat to moist areas beneath logs and other ground litter; individuals have even been found hiding in cracks in the mud created by the drying of a pond. With their strong legs, they are capable of jumping as much as 3 feet vertically or horizontally. They are known to travel long distances overland, with juveniles moving more than 300 feet from wetland habitats and adults moving more than 0.8 mile between aquatic sites. Cricket frog tadpoles usually seek shelter in aquatic vegetation.

Most northern cricket frogs have a dark triangle between their eyes.

Northern cricket frogs can be various shades of brown and green.

FOOD AND FEEDING Northern cricket frogs eat small terrestrial or flying invertebrates such as spiders, worms, ants, beetles, and flies. They forage both at night and during the day. The tadpoles eat algae and microscopic plants.

DESCRIPTION OF CALL The advertisement call is a series of rapidly produced clicks that sound like someone striking two marbles together.

REPRODUCTION The breeding period in the Southeast typically ranges from early spring through the summer and may extend into autumn if temperatures remain warm, and takes place in a variety of aquatic habitats. Females lay from one to several eggs at a time, totaling as many as 400. The eggs may be attached to aquatic vegetation, may float on the water's surface, or may rest on the bottom. After the eggs hatch, the tadpoles can take as few as 4 and as many as 12 weeks to metamorphose.

PREDATORS AND DEFENSE Northern cricket frogs have numerous vertebrate predators, including bullfrogs, bass, garter snakes, watersnakes, turtles, kestrels, grackles, and carnivorous mammals. They avoid predation by jumping erratically on land, across the top of aquatic vegetation, or into the water. Tadpoles subject to predation by dragonfly larvae (naiads) develop a black tail tip that may act as a "false head" to divert the naiad's strike. In aquatic situations where fish are prevalent, the tadpoles have a transparent tail, presumably to make them less conspicuous.

CONSERVATION The northern cricket frog is not recognized as a species in need of conservation in the Southeast. Severe population declines have been noted in portions of the Midwest, Canada, and as far south as West Virginia; the reasons for these declines have not been determined. The species is listed as Endangered, Threatened, or of Special Concern in West Virginia, Indiana, Illinois, Michigan, Minnesota, and Wisconsin.

COMMENTS The northern cricket frog has three subspecies, and two occur in the Southeast. The eastern cricket frog (*Acris crepitans crepitans*) is found in all of the southeastern states; Blanchard's cricket frog (*A. c. blanchardi*) is found in most of northern Tennessee and the remainder of the geographic range to the north and west. The subspecies differ from each other in basic body color and skin texture, Blanchard's cricket frog usually being solid brown or gray with more granular skin.

Southern cricket frogs are common near the edges of lakes, ponds, and rivers.

Southern Cricket Frog

Acris gryllus

DESCRIPTION Southern cricket frogs are small and colorful. The body color may be a mixture of brown, green, and gray and may have overlying markings of yellow, orange, red, or black. Some individuals are a uniform color. The belly is grayish or white. Small but apparent warts cover the head, body, and legs. The maximum body length of adults is 1.25 inches. The head is relatively pointed. Like other members of the genus, the southern cricket frog has a backward-pointing dark triangle between the eyes. A well-defined, smooth-edged dark stripe runs along the back of the thigh; the subspecies found in Florida (Florida cricket frog, *A. g. dorsalis*) has two dark stripes instead of one. Partial webbing connects the first and second toes of the hind feet.

WHAT DO THE TADPOLES LOOK LIKE? Southern cricket frog tadpoles are light to dark greenish brown with a long, translucent tail that can have dark spots or speckling. The tip of the tail is often solid black. The top of the tail musculature may have dark bars or spots. The longest southern cricket frog tadpoles are typically less than 1.5 inches.

SIMILAR SPECIES The southern cricket frog can be distinguished from other frogs by its small size and warty skin, the dark triangle on the head, and the presence of one or two smooth-edged dark stripes on the rear of

How do you identify a southern cricket frog?

SKIN
Warty

LEGS
Muscular and long

FEET AND TOES
Hind feet webbed

BODY PATTERN AND COLOR
Usually a combination of brown, gray, green, yellow, orange, red, and/or black

DISTINCTIVE CHARACTERS
Dark triangle between eyes; hind leg with one or two well-defined stripes

CALL
Steady series of single clicks

SIZE
max tadpole = 1.75"
typical adult = 1"

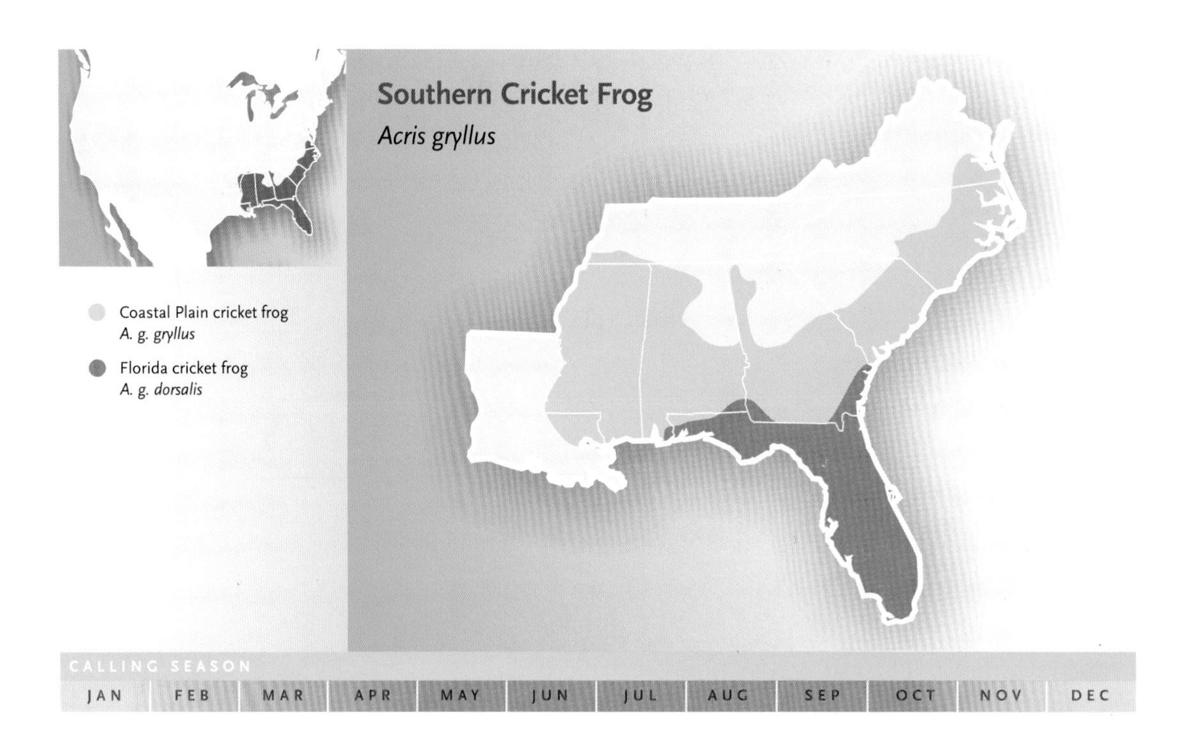

Southern cricket frogs have pointed heads.

the thighs. This species has less webbing on the hind feet, a more pointed head, and a less jagged border on the leg stripe than the northern cricket frog. Southern and northern cricket frogs occasionally interbreed in areas where both occur and can produce hybrids that are difficult to identify.

DISTRIBUTION AND HABITAT Southern cricket frogs are found in all or part of every southeastern state, including all of Mississippi and Florida. They are restricted to the Coastal Plain in the Carolinas and Virginia, to a few southwestern areas in Tennessee, and to the eastern edge of Louisiana. They are present in most of Georgia and Alabama, although they are rare outside the Coastal Plain. Southern cricket frogs are found in habitats associated with aquatic areas, including those in uplands, bottomlands, and stream margins. They are commonly seen on land or in vegetated wet areas, particularly in open, grassy habitats or those with low-lying shrubs at the edges of wetlands such as Carolina bays, farm ponds, and swamps.

BEHAVIOR AND ACTIVITY Southern cricket frogs are active year-round on warm days and can be seen both day and night in open areas around aquatic habitats, although they are noticeably less active during winter. Despite their small size, they can jump more than 3 feet vertically or horizontally. Unlike treefrogs, cricket frogs seldom climb into bushes or trees. During periods of drought or extreme cold, southern cricket frogs seek refuge under ground litter such as logs, dead leaves, or other vegetation. The tadpoles characteristically hide in aquatic vegetation.

A southern cricket frog from Aiken, South Carolina

FOOD AND FEEDING Adults feed on land and eat small invertebrates such as springtails, spiders, flies, beetles, and other small flying insects. The tadpoles are presumed to eat algae attached to the surface of aquatic vegetation.

DESCRIPTION OF CALL The advertisement call is a steady series of clicks produced singly but increasing in speed as an individual continues to call. The sound resembles that made by hitting two marbles together. Some people actually hit marbles together to elicit calling in male cricket frogs.

REPRODUCTION Breeding can occur in any month in the Southeast, especially in Florida, although breeding

A male southern cricket frog calling to attract a mate

activity typically peaks from April to October. Virtually any aquatic habitat may be used. Females lay 1–10 eggs at a time, totaling as many as 250. The eggs are sometimes attached to aquatic vegetation or may be on the bottom in shallow water. They hatch in 4 days or more; after as few as 6 weeks or as many as 13, the tadpoles metamorphose into froglets that are about 0.3–0.5 inch long.

PREDATORS AND DEFENSE Several varieties of snakes (watersnakes, garter snakes, and pine woods snakes) and fish (redfin pickerel, bluegill, and large-mouth bass) are known predators, and other frogs, toads, and birds probably eat southern cricket frogs as well. When threatened, these frogs will jump around on land or into the water, where they swim to the bottom.

Did you know?

The world's smallest frogs—the Brazilian gold frog and the Cuban mini-frog—are only three-eighths of an inch long, smaller than any other land-dwelling vertebrates. Two of them could sit end to end on a nickel without touching the edges.

CONSERVATION Degradation of terrestrial habitat surrounding wetlands, including certain site preparation activities associated with commercial pine plantations, can have a major negative impact on local populations of southern cricket frogs, but in the early 2000s the species remained widespread and abundant in most of the Southeast.

Some southern cricket frogs are almost entirely green.

COMMENTS Two subspecies have been described: the Coastal Plain cricket frog (*Acris gryllus gryllus*) and the Florida cricket frog (*A. g. dorsalis*). The Florida subspecies occurs throughout Florida and in adjacent areas of Alabama and Georgia; the Coastal Plain cricket frog occupies the remainder of the geographic range. The Coastal Plain cricket frog has a single dark line behind the thigh while the Florida cricket frog has two lines.

An upland chorus frog from South Carolina

How do you identify an upland chorus frog?

SKIN
Granular

LEGS
Long and slender

FEET AND TOES
Limited webbing on hind feet

BODY PATTERN AND COLOR
Gray or tan, stripes on back may be broken into long spots

DISTINCTIVE CHARACTERS
Light stripe above lip and pointed snout

CALL
Ascending trill or *crreeeek* like the sound of a finger running over a comb

SIZE
max tadpole = 1.25"
typical adult = 1"

Upland Chorus Frog *Pseudacris feriarum*

DESCRIPTION These small, slender frogs have relatively long legs and little webbing between the rear toes. The body is gray, tan, or brown, and the background color becomes lighter in active frogs. Adults have a prominent light stripe on the upper lip and a fairly wide, dark band running from each nostril through the eye and down each side. They generally have three longitudinal dark stripes down the back, but the stripes may be broken up into long spots or may even be missing altogether. A triangular spot is often present between the eyes. The belly is white or cream with some dark spotting on the chest.

WHAT DO THE TADPOLES LOOK LIKE? The tadpole is relatively small and dark brown on the back. The belly is lighter, and there is usually a dark stripe running the length of the musculature in the tail. The tail fin is transparent but may have some dark stippling.

SIMILAR SPECIES Upland chorus frogs vary in color pattern and are easily confused with southern chorus frogs and Brimley's chorus frogs. They can usually be distinguished on the basis of geographic range. In addition, Brimley's chorus frogs have longitudinal bars on the upper rear legs while upland chorus frogs have transverse bars. Southern chorus frogs have a more pointed snout than upland chorus frogs and tend to have more

Some upland chorus frogs lack markings on the back.

black markings and more granular skin.

DISTRIBUTION AND HABITAT Upland chorus frogs are part of a group of very similar species that range across most of the eastern states and even into the western United States. The boundaries between the different species are not well understood. The upland chorus frog's range includes Louisiana eastward through most of Mississippi and Alabama and through the northern two-thirds of Georgia. The range continues northward through the Carolinas, primarily in the Piedmont and in parts of the Coastal Plain, and includes most of Virginia. It is found in a few locations in the Florida panhandle. See "Comments" (below) regarding the taxonomy and our understanding of the geographic ranges of the chorus frogs. Upland chorus frogs are typically abundant in shallow bodies of water with extensive grasses and other emergent vegetation, but they can be found in almost any water body, including slow streams, farm ponds, Carolina bays, river-bottom swamps, bogs, marshes, and grassy swales.

BEHAVIOR AND ACTIVITY Upland chorus frogs are most active in either winter or early spring. They are rarely found on the surface during the remainder of the year and apparently burrow or hide under surface objects to avoid heat and desiccation. They are often active during the daytime, especially in late afternoon, and sometimes continue to call even when the temperature drops close to freezing.

FOOD AND FEEDING This species, like its close relatives, probably feeds exclusively on small insects and other arthropods such as ants, small beetles, and flies. The tadpoles scrape algae and detritus from grass stems and other underwater objects.

DESCRIPTION OF CALL The call is best described as the sound of a thumb running over the teeth of a comb. The ascending *crreeeekk* is usually faster than that of the southern chorus frog and slower than that of Brimley's chorus frog, with which it can be easily confused. The easiest way to distinguish the call of the southern chorus frog from the upland chorus frog is to try to count the pulses within the call. If the pulses can be counted by ear (usually 6–12 pulses), it is a southern chorus frog. If the pulses are produced too quickly to count (usually 14–35 pulses), it is an upland chorus frog.

A pair of upland chorus frogs in amplexus

REPRODUCTION Females may lay more than 500 eggs during a single breeding season. Eggs are laid in small clusters attached to grass stems just below the surface. They hatch within a few days, longer if the water temperature is low. The tadpoles transform into small frogs within about 10 weeks but probably stay close to or within the wetland for some time afterward.

PREDATORS AND DEFENSE Predators include aquatic invertebrates such as large diving beetles and fishing spiders along with various species of birds, including owls. Garter snakes and ribbon snakes likely eat large numbers of these frogs if the weather is warm enough for the snakes to be active. Other likely predators include watersnakes of various species, turtles, and small to medium-sized mammals.

CONSERVATION The upland chorus frog is a species of conservation concern in Pennsylvania and West Virginia; its status throughout the rest of

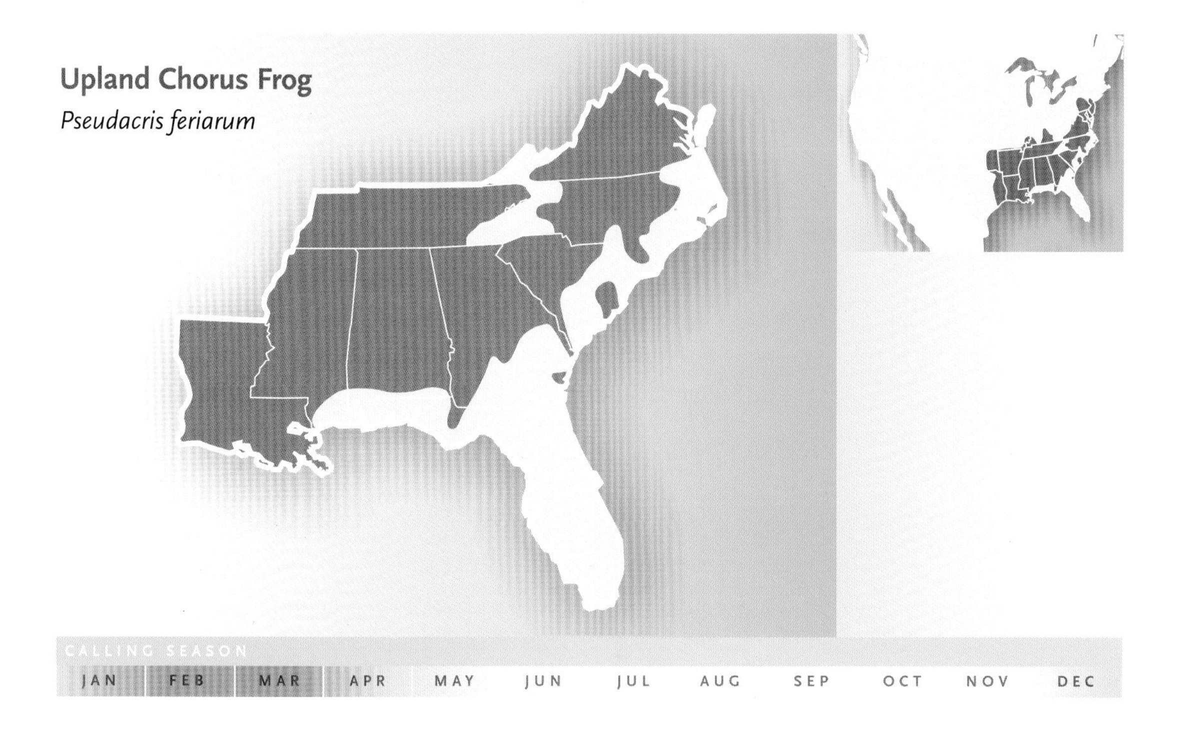

Upland chorus frogs vary in appearance geographically.

Upland chorus frogs are common residents of the Piedmont.

its range has not been studied. Other species within the chorus frog species "complex" are imperiled in other parts of the United States. Human activities have a detrimental effect on upland chorus frogs, but these frogs can be found near and within moderately developed areas.

COMMENTS Once known as *P. triseriata*, this species is part of a group of species and subspecies of chorus frogs found throughout much of the United States that is often referred to as the "*triseriata* complex." It is difficult to distinguish among them, especially on morphological examination alone. Research has shown that upland chorus frogs from western Mississippi, Louisiana, Arkansas, and eastern Texas and Oklahoma probably represent a distinct species described in 2008 as the Cajun chorus frog, *P. fouquettei*; and that chorus frogs in the Delmarva Peninsula should be considered a full species as well.

The relatively smooth skin of the southern chorus frog is covered with small, granular bumps.

How do you identify a southern chorus frog?

SKIN
Finely granular

LEGS
Long and slender

FEET AND TOES
Limited webbing on hind feet

BODY PATTERN AND COLOR
Gray or tan; spots on back in three rows sometimes forming stripes

DISTINCTIVE CHARACTERS
Light stripe above lip; pointed snout

CALL
Slow, harsh trill somewhat like a ratchet

SIZE
max tadpole = 1.5"
typical adult = 1"

Southern Chorus Frog *Pseudacris nigrita*

DESCRIPTION These small, slender frogs are typically gray or tan with well-defined spots on the back in three longitudinal rows. In many individuals, the spots join to form three broken stripes. Individuals may vary in the shade of their background color from light gray to nearly black depending on temperature, time of day, and activity. A prominent white line runs along the upper lip, and a dark mask passes through each eye and continues as a dark band along each side. The skin is relatively smooth but is covered with small, granular bumps. The rear legs have dark transverse bars and the belly is whitish. The throat of males is darker than that of females and is often yellowish or olive. Individuals in the northern part of the range (Virginia and North Carolina) are mostly some shade of brown; individuals in southern areas (South Carolina, Georgia, and Florida) have more black markings.

A dark stripe through each eye helps to distinguish the southern chorus frog from other species.

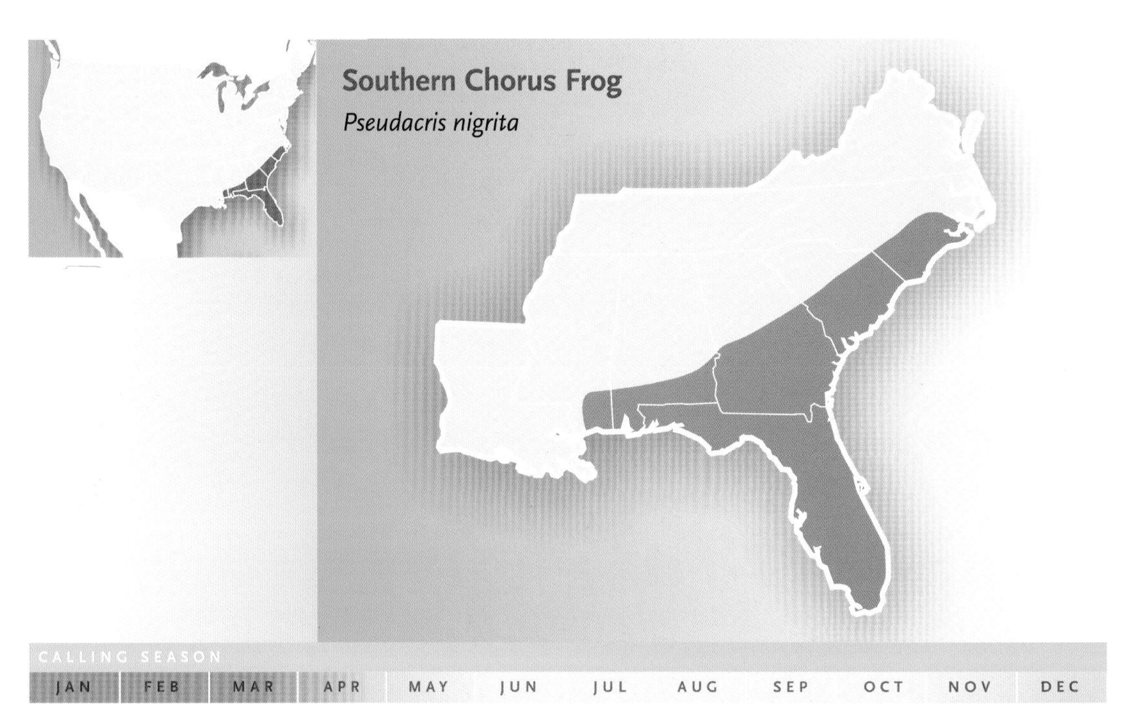

A southern chorus frog tadpole

WHAT DO THE TADPOLES LOOK LIKE? The tadpoles are dark brown or copper colored and are covered with small, light spots. Dark longitudinal bands and spots extend onto the musculature of the tail. The tail fin is transparent with little spotting. The tadpoles undergo metamorphosis at a length of about 1.5 inches.

SIMILAR SPECIES Southern chorus frogs are most easily confused with the two similar species of chorus frogs. They can be distinguished from the upland chorus frog on the basis of geographic range and head shape. The southern chorus frog has a narrower, more pointed snout and more black markings than the upland chorus frog. Southern chorus frogs have transverse bars on the upper part of their rear legs while Brimley's chorus frogs have longitudinal bars.

DISTRIBUTION AND HABITAT Southern chorus frogs range from the Coastal Plain in North Carolina southward through all of Florida and west to extreme

Male southern chorus frogs call in winter and early spring to attract mates.

southeastern Louisiana. These frogs are associated with pine forests and call in grassy wet areas in or near these habitats. They are often found in Carolina bays and man-made habitats such as borrow pits and roadside ditches.

BEHAVIOR AND ACTIVITY Southern chorus frogs are rather cold tolerant and can be quite active under conditions in which most other frogs become dormant. They are active from December through April in most parts of the Southeast and often call during the day, frequently starting in early afternoon and continuing after dark until it becomes too cold. During periods of heavy rain—for example, during hurricanes—they may call even during the summer or fall, although it is unlikely that they breed. Calling males are difficult to locate because they often vocalize from within clumps of grass near the surface of the water. When not calling, southern chorus frogs are rarely found and apparently spend most of their time in burrows or under logs or other objects.

FOOD AND FEEDING Like most other small frogs, southern chorus frogs feed almost exclusively on small insects. Stomach contents of preserved specimens have included primarily ants and small beetles along with grasshopper nymphs, which they apparently capture during the daytime in tall grass. Tadpoles feed on algae and detritus on subsurface and bottom objects.

DESCRIPTION OF CALL The call, which is relatively loud for such a small frog, can best be described as a slow trill. To some people it resembles the sound of a ratchet. There are typically 6–12 notes per trill, and the trills are typically separated by at least several seconds. The number of pulses in the call of a southern chorus frog can usually be counted by ear, while those of similar chorus frog species are produced too rapidly to count.

REPRODUCTION Throughout most of the Southeast, the breeding season begins in December or January and typically continues through March or April. The peak breeding period in southern Florida is late winter to early spring. The 150 or so eggs each female produces are laid in small clusters of 10–20 that are attached to vegetation or other subsurface objects. The eggs hatch within about 3 days, and the tadpoles reach a length of about 1.5 inches before they transform into froglets some 1.5–4 months later. The young frogs take several months to mature and are likely to remain near the wetland for some time.

PREDATORS AND DEFENSE Predators probably include several species of wading birds, owls, small to medium-sized mammals such as raccoons, predatory insects, and larger frogs. Research has shown that mortality is fairly high, and individuals that do reach maturity are likely to only survive one breeding season. Tadpoles are eaten by numerous species of aquatic salamander larvae and predatory insects. These frogs may avoid predation to some degree by breeding during cold months when some potential predators (e.g., aquatic snakes) are less active and by calling while hidden within dense clumps of grass.

Southern chorus frogs are often covered in tiny bumps.

CONSERVATION Southern chorus frogs are fairly abundant throughout their range, but populations have probably declined in areas with large-scale silviculture, fire suppression, and high levels of urbanization. Southern chorus frogs are not listed as Endangered or Threatened in any state in which they are found.

COMMENTS A subspecies known as the Florida chorus frog (*P. n. verrucosa*) is no longer recognized. The taxonomic relationships among several of the chorus frogs are unclear, but taxonomists are currently proposing new relationships that should help to clarify the species designations within this complex group.

Brimley's chorus frogs are named after C. S. Brimley, a North Carolina naturalist from the early 1900s.

Brimley's Chorus Frog *Pseudacris brimleyi*

DESCRIPTION These small, slender, long-legged chorus frogs are primarily tan or brown. Adults always have a well-defined dark stripe that runs unbroken from the snout through each eye and along each side to the groin. They may have lighter stripes on the back as well, but these are often missing or indistinct. The belly is usually yellow or cream and the chest is often peppered with spots. The markings on the legs tend to run lengthwise rather than forming bands as those on many other chorus frogs do.

WHAT DO THE TADPOLES LOOK LIKE? The tadpoles are dark brown; the top part of the tail musculature is dark and the lower half is light tan. The throat is dark and the tail fin is transparent. The tadpoles reach a length of about 1.5 inches before they undergo metamorphosis.

SIMILAR SPECIES Brimley's chorus frogs may be confused with southern chorus frogs and upland chorus frogs but are relatively easily distinguished by the bold, dark band along each side; the lack of a triangle between the eyes; and the lengthwise bands on the legs. Froglets might be confused with little grass frogs, but adults get considerably larger. Ornate chorus frogs are stouter and usually have bolder markings on the back.

How do you identify a Brimley's chorus frog?

SKIN
Smooth with tiny bumps

LEGS
Long and slender

FEET AND TOES
Limited webbing on hind feet; long toes

BODY PATTERN AND COLOR
Tan or brown; bold stripe from snout through eye and along each side

DISTINCTIVE CHARACTERS
Dark stripe along sides; markings on legs run lengthwise

CALL
Short, raspy trill repeated rapidly

SIZE
max tadpole = 1.5"
typical adult = 1"

Brimley's chorus frogs breed in shallow wetlands.

DISTRIBUTION AND HABITAT Brimley's chorus frogs are found in the Coastal Plain from Virginia southward through the Carolinas to extreme eastern Georgia just south of the Savannah River. These frogs inhabit forested or open areas and adjacent marshes, shallow flooded fields, open grassy swales, roadside ditches, and the margins of swamps. A preferred breeding habitat at one South Carolina site is a seasonally flooded hardwood swamp dominated by laurel oak and dwarf palmetto that occurs adjacent to small streams in the Savannah River floodplain.

BEHAVIOR AND ACTIVITY Brimley's chorus frogs are active from late winter into March or April. When not breeding they can be found some distance from water, but they are rarely seen outside the breeding season. They often call during the daytime, especially in the afternoon, and will call after dark for several hours as well. They probably spend the coldest periods of the winter underground or under logs.

FOOD AND FEEDING Little is known about their diet, but these frogs presumably feed on a variety of small invertebrates including ants, small beetles, spiders, and flies. The tadpoles presumably eat detritus and algae present in the shallow pools where they are developing.

DESCRIPTION OF CALL The call is a short, raspy, buzzlike trill. It is shorter than the call of southern chorus frogs—less than a second in duration—and the individual notes are less distinct and are produced much more rapidly.

REPRODUCTION Brimley's chorus frogs begin breeding in late winter and continue through early spring. Each female lays her 300 or so eggs in small clusters attached to stems or other objects. The eggs hatch in a few days and the tadpoles take about 1–2 months to transform. The tadpoles reach a length of about 1.5 inches before they transform into froglets.

A Brimley's chorus frog (top right) from near Columbia, South Carolina

The presence of three well-defined stripes on the back (bottom right) distinguishes the Brimley's chorus frog from other species of chorus frogs.

PREDATORS AND DEFENSE Eastern ribbon snakes are known predators, and garter snakes and several species of watersnakes may prey on these frogs as well. Large, predatory aquatic insects and insect larvae likely take the tadpoles and may even consume adult frogs.

CONSERVATION Brimley's chorus frog populations are apparently stable throughout most of the species' range, but are listed as a Species of Concern in Georgia. These frogs may be vulnerable to intense forestry practices that alter the wetlands they depend on for breeding. Urbanization almost assuredly has resulted in the local extinction of populations in many areas throughout the range.

COMMENTS Brimley's chorus frog is named after Clement Samuel Brimley, a North Carolina naturalist who studied amphibians and reptiles in the early 1900s.

Brimley's Chorus Frog

Pseudacris brimleyi

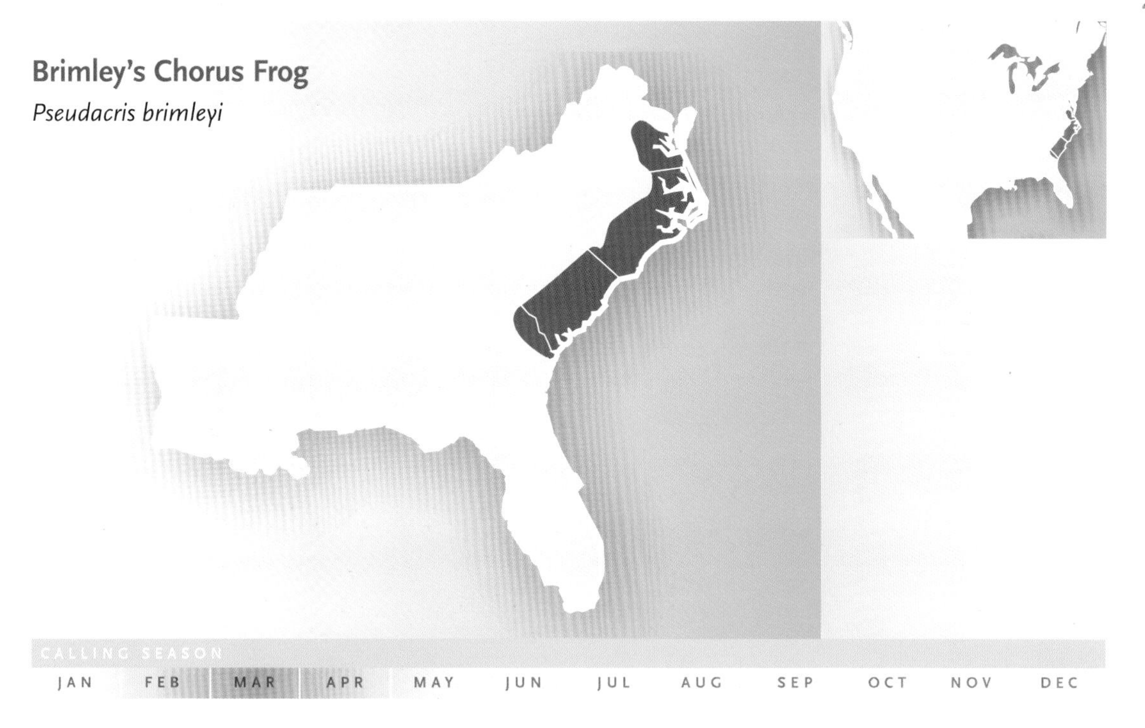

CALLING SEASON

JAN FEB MAR APR MAY JUN JUL AUG SEP OCT NOV DEC

Little grass frogs are agile climbers.

How do you identify a little grass frog?

SKIN
Smooth

LEGS
Long and slender

FEET AND TOES
Moderate webbing on hind feet

BODY PATTERN AND COLOR
Brown or tan, sometimes reddish

DISTINCTIVE CHARACTERS
Tiny size; pointed snout; dark stripe from snout along each side to hind legs

CALL
Very high-pitched, tinkling trill

SIZE
max tadpole = 1"
typical adult = 0.5"

Little Grass Frog

Pseudacris ocularis

DESCRIPTION The little grass frog is the smallest frog in the United States; adults rarely exceed 0.5 inch in length. They are slender in build with relatively long legs and a pointed snout. Adults are generally some shade of brown but may range from light tan to dark brown to nearly reddish. Some individuals even have a greenish cast. Their color varies with activity and temperature—they tend to be lightest on warm nights and darkest when the weather is colder. A wide, dark brown stripe usually runs from the snout through each eye and along each side. Just behind the eye, the stripe is usually bordered below by white. A triangular dark spot is often present between the eyes, and the back may have relatively obscure markings. The beadlike eyes are positioned more on the sides of the head than in most other frogs. They have tiny, barely recognizable toe pads. During the breeding season, calling males have a large vocal sac and a dark throat.

WHAT DO THE TADPOLES LOOK LIKE? The tadpoles are pinkish or coppery and sometimes have an olive back. The central, mus-

The little grass frog is the smallest North American frog.

cular area of the tail has a dark, wide stripe, and the relatively translucent tail fin has dark spots. Although they are longer than the adult form, tadpoles reach a total length of only about 1 inch.

SIMILAR SPECIES Little grass frogs are most likely to be confused with other species of chorus frogs or with cricket frogs, but their very small size, relatively smooth skin, delicate body, pointed nose, and lateral stripe make them relatively easy to distinguish from other closely related species within their range.

DISTRIBUTION AND HABITAT Little grass frogs are found primarily in the Coastal Plain from southeastern Virginia southward through all of peninsular Florida and in the Florida panhandle east of Choctawhatchee Bay. These tiny frogs typically inhabit grassy, shallow, open wetlands; roadside ditches; and small Carolina bays. They can be extremely abundant in wet savannas and wetlands in open pine flatwoods. They are sometimes found along the margins of river swamps and cypress ponds if sufficient emergent vegetation is present. They can be common in areas where electrical power line rights-of-way bisect forests that include shallow wetland habitats.

BEHAVIOR AND ACTIVITY Little grass frogs are active during the day and at night. They may be active year-round in Florida and throughout all but the coldest months in the rest of their range. They often hide near the ground

The call of a little grass frog is so high-pitched that some people cannot hear it.

Little Grass Frog

Pseudacris ocularis

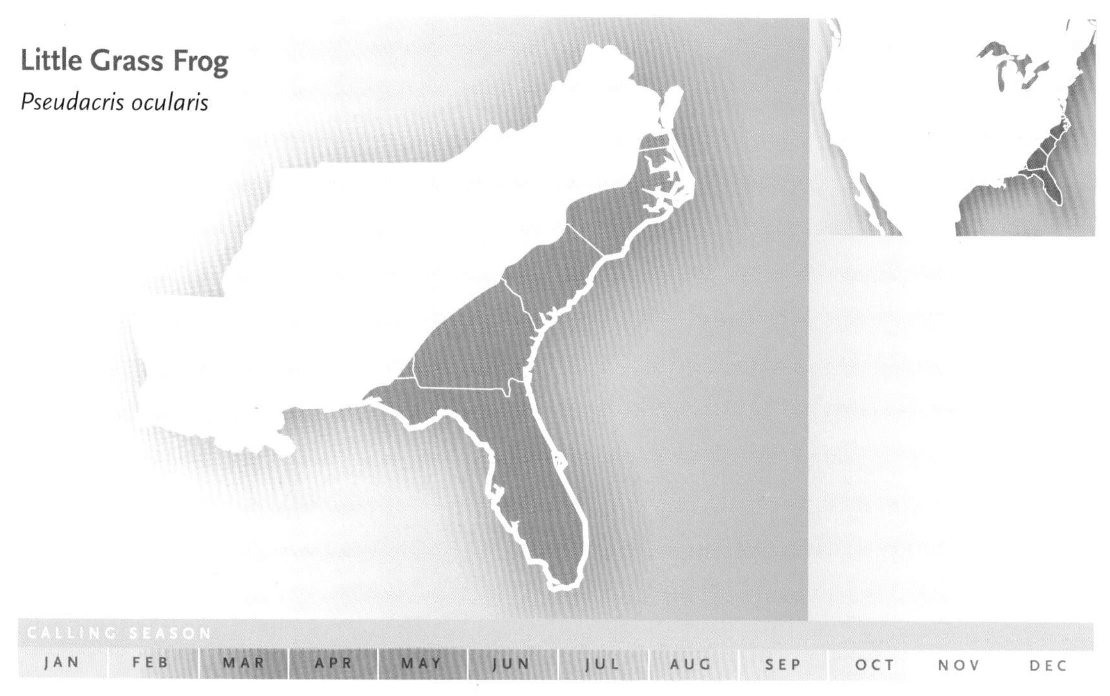

among thick grasses and sedges, and even though a chorus can be heard calling, individual frogs are often difficult to locate. They also call from tall grass stems and vines, and climb into small bushes.

A little grass frog from South Carolina

FOOD AND FEEDING Little grass frogs eat a variety of small insects, including ants, leafhoppers, springtails, and small beetles, and are also known to eat mites. Most of the prey items that have been recorded are terrestrial, indicating that the frogs probably forage on the ground away from the water. The tadpoles use their mouthparts to scrape algae and detritus from aquatic vegetation and other underwater surfaces.

DESCRIPTION OF CALL The advertisement call is a very high-pitched, tinkling trill that can easily be mistaken for that of a small insect such as a cricket. The call is so high in pitch that some humans are unable to hear it.

REPRODUCTION The males usually call from perches on grass stems slightly above the water's surface. The breeding season generally peaks in March–May, but males may call at any time of year, especially in southern locales. Eggs are laid either individually or in small clusters on the bottom of wetlands or attached to vegetation. The eggs hatch in 1–2 days and the tadpoles take about 2 months to transform into froglets that are about a third of an inch long.

PREDATORS AND DEFENSE Tadpoles and adults likely have a variety of aquatic and semiaquatic predators, including snakes, turtles, wading birds, and predatory insects. Wolf spiders have been known to capture and eat the adults. These frogs generally rely on their small size and cryptic coloration to conceal them from predators. Some researchers have suggested that the lateral stripe on the sides may help camouflage the frogs when they rest vertically on grass stems. When disturbed, they can jump more than 1.5 feet, a surprisingly long distance for their size.

CONSERVATION Little grass frogs can be abundant in appropriate habitat and seem able to tolerate moderate human disturbance because they breed in roadside ditches and power line rights-of-way. No declines in their populations have been noted, and they appear to be in no need of special conservation efforts.

COMMENTS Amphibian biologists long debated the relationship of the little grass frog to other species in the treefrog family, assigning it to more than a half-dozen different genera until molecular genetics analyses in the 1980s demonstrated the species' affiliation with the genus *Pseudacris*.

Most ornate chorus frogs are some shade of gray, and all generally have bold markings.

Ornate Chorus Frog *Pseudacris ornata*

DESCRIPTION The ornate chorus frog is the largest of the southeastern chorus frogs and one of the most strikingly patterned frogs in the Southeast. The relatively stout adults are characterized by well-defined dark markings on a brown, reddish brown, gray, or—rarely—bright green background. A dark stripe passes through each eye and along each side, although it may break up toward the rear of the body. Most of the markings are lined in white. The forelimbs and hind limbs are marked with dark bars, and there is some yellow in the groin area. Most individuals have a dark triangle on top of the head between the eyes. Some individuals have few or no markings on the back.

WHAT DO THE TADPOLES LOOK LIKE? The tadpoles are uniformly dark, with light spotting in some individuals. The throat is usually white. Young tadpoles have a bicolored tail. The tail fin is rather high, nontransparent, and dark like the tadpole's body. Tadpoles reach a length of about 1.75 inches before they transform.

Ornate chorus frogs are quite variable in color.

How do you identify an ornate chorus frog?

SKIN
Smooth

LEGS
Moderately long and somewhat stout

FEET AND TOES
Limited webbing on hind feet

BODY PATTERN AND COLOR
Brown or gray, sometimes green; bold markings on back

DISTINCTIVE CHARACTERS
Dark stripe through eye and along each side; yellow in groin area

CALL
Repetitive metallic, monotonic peep

SIZE
max tadpole = 1.75"
typical adult = 1"

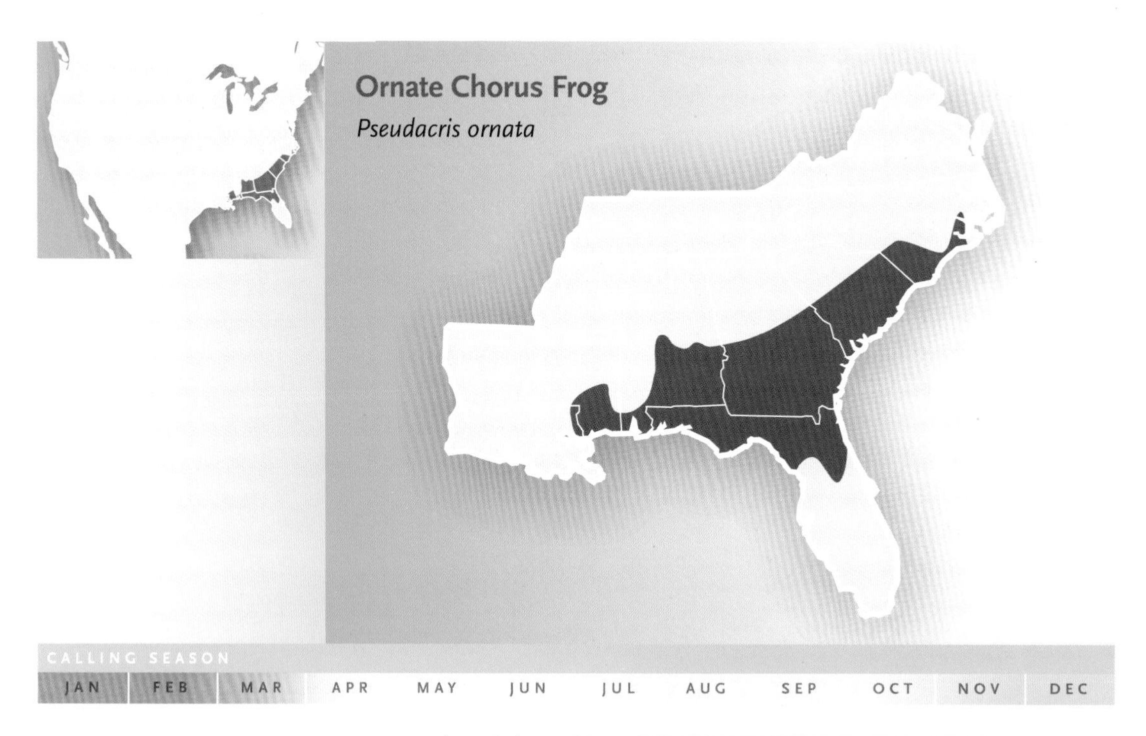

The red phase of the ornate chorus frog

SIMILAR SPECIES Their stout body differentiates ornate chorus frogs from the true frogs and toads. They can be distinguished from other chorus frogs whose range overlaps theirs by their stout build, bold face mask, yellowish groin, and distinct blotches on the back. They can be distinguished from all of the treefrogs by their lack of enlarged toe pads.

DISTRIBUTION AND HABITAT Ornate chorus frogs are animals of the southeastern Coastal Plain. They are found from North Carolina, where they range into the sandhills, south through the northern half of Florida and westward through the lower halves of Georgia and Alabama to extreme eastern Louisiana. Ornate chorus frogs are found in association with various types of aquatic habitats within or adjacent to forests. They most commonly breed in ephemeral wetlands that hold water during the winter and spring and dry out during the hot summer months; these include cypress ponds, wet meadows, roadside ditches, and Carolina bays. During the summer and fall, these frogs apparently retreat into shallow burrows or hide under objects to avoid desiccation.

BEHAVIOR AND ACTIVITY Ornate chorus frogs are most active during some of the coldest months. They are often the first species to begin calling in early winter (or even early fall) and often end their breeding activities just as the weather begins to warm in the early spring. Males may call at near-freezing temperatures, and adults have been seen basking in the sun to warm up when the ground was partially covered with snow. The males often call while sitting in shallow water or even floating on the surface. During the remainder of the year they are nearly impossible to find because they remain hidden in shallow burrows within the surrounding forest, although very heavy rains may bring them to the surface even during warm months.

The green phase of the ornate chorus frog

Bold facial stripes and a stocky build distinguish the ornate chorus frog from other chorus frogs.

FOOD AND FEEDING Ornate chorus frogs feed primarily on small insects and other invertebrates and may eat earthworms and insect larvae while underground in burrows. They may burrow under or adjacent to grass and other vegetation to prey on insect larvae attracted to the roots and on adult insects attracted to the stem or leaves. Tadpoles likely feed on algae and other detritus within their wetland.

DESCRIPTION OF CALL The advertisement call is best described as a loud, short whistle repeated about 60–80 times per minute. The call is most similar to a spring peeper's but differs in that it terminates abruptly and

Ornate chorus frogs are most active during the winter.

Developing ornate chorus frog eggs

does not increase in pitch toward the end of each whistle. Some observers liken the call to a hammer striking steel.

REPRODUCTION Males may begin calling in late fall, but actual breeding usually does not take place until later in December or January and continues into March. Eggs are laid in clusters of 20–100 and are usually attached to aquatic vegetation. The eggs hatch within a few days and the tadpoles transform into froglets within about 3 months. The tadpoles can reach a length of about 1.75 inches before they metamorphose.

PREDATORS AND DEFENSE Potential predators of tadpoles include salamander larvae, dragonfly larvae, and other aquatic invertebrates such as predaceous water beetles. Some of these predators probably also prey on adults when they are in aquatic habitats. The southern hognose snake is the only recorded predator of adults, but it is likely that other snakes, wading birds, and small to medium-sized mammals eat them, too. Ornate chorus frogs are not known to be particularly toxic, but by confining their activity to the coldest parts of the year they may avoid predation by many reptiles (e.g., watersnakes) and migratory wading birds.

CONSERVATION Researchers have documented apparent declines of ornate chorus frog populations in areas with high-intensity forestry practices that disturb the soil or alter wetland breeding sites. Alteration of naturally occurring fire regimes and development of industrial pine plantations may cause a substantial loss of populations in many areas. This species is not legally protected by any state or federal regulations.

Strecker's chorus frogs are extremely stout compared with most other species of chorus frogs.

How do you identify a Strecker's chorus frog?

SKIN
Smooth

LEGS
Stocky

FEET AND TOES
Limited webbing on hind feet

BODY PATTERN AND COLOR
Gray or tan, usually spotted

DISTINCTIVE CHARACTERS
Dark stripe through and behind eye; dark spot below eye

CALL
Rapid, high-pitched, bell-like whistle

SIZE
max tadpole = 1.5"
typical adult = 1.5"

Strecker's Chorus Frog *Pseudacris streckeri*

DESCRIPTION These stout chorus frogs have a well-defined dark band that runs from the snout through each eye to the armpit. In some individuals, the dark band continues along each side. The background coloration is usually tan but may be more grayish or may even have a greenish tinge. Strecker's chorus frogs usually have well-defined, though sometimes faint, spots on the back; these may be darker brown but are often green. Some individuals lack spots on the back entirely. A dark spot is present under each eye, and bands are present on the hind legs. The belly is usually white or cream and the groin is yellow or orange.

WHAT DO THE TADPOLES LOOK LIKE? The tadpole is relatively small and robust. The back is black or brown with limited mottling; the belly is white. Tadpoles reach approximately 1.5 inches before they transform into froglets.

The tadpole of the Strecker's chorus frog is relatively small and robust.

SIMILAR SPECIES Strecker's chorus frogs can be distinguished from all other chorus frogs within their range by the lack of a white line on the upper lip. Their stout body form might cause them to be confused with small toads, but close inspection reveals the distinctive bold face mask.

DISTRIBUTION AND HABITAT Strecker's chorus frogs are found throughout the eastern half of Texas, all but the panhandle of Oklahoma, and in scattered localities in Arkansas and the Midwest. In the Southeast, they are found only at a few sites in northwestern Louisiana and in one parish in southwestern Louisiana. They occupy a variety of habitats, including moist woodlands, stream borders, open prairies, and even agricultural areas.

Louisiana is the only southeastern state where Strecker's chorus frogs can be found.

BEHAVIOR AND ACTIVITY Strecker's chorus frogs are active in the Southeast primarily in winter and early spring. The remainder of the year they spend underground, although they may emerge at any time during very heavy rains. They are mostly nocturnal and move from their burrows to breeding sites primarily at night. Males do sometimes call during the day, especially when it is cloudy and rainy. Adults use their stout forelimbs to burrow into loose soil headfirst, rather than backward as toads do. Adults may live several years and have relatively high survivorship once they reach adulthood.

The spots on some Strecker's chorus frogs are green.

FOOD AND FEEDING Strecker's chorus frogs probably eat a wide variety of insects and other arthropods; prey items recorded include leafhoppers, beetles, flies, moths, and ants. Studies have shown that they can feed underground and may eat beetle larvae and earthworms they encounter within their burrows. The tadpoles eat algae and may be cannibalistic if food is scarce.

DESCRIPTION OF CALL The call is a clear, high-pitched whistle somewhat similar to that of the ornate chorus frog. The note is repeated several times, and large choruses have been described as sounding like a squeaky wagon wheel.

Strecker's Chorus Frog

Pseudacris streckeri

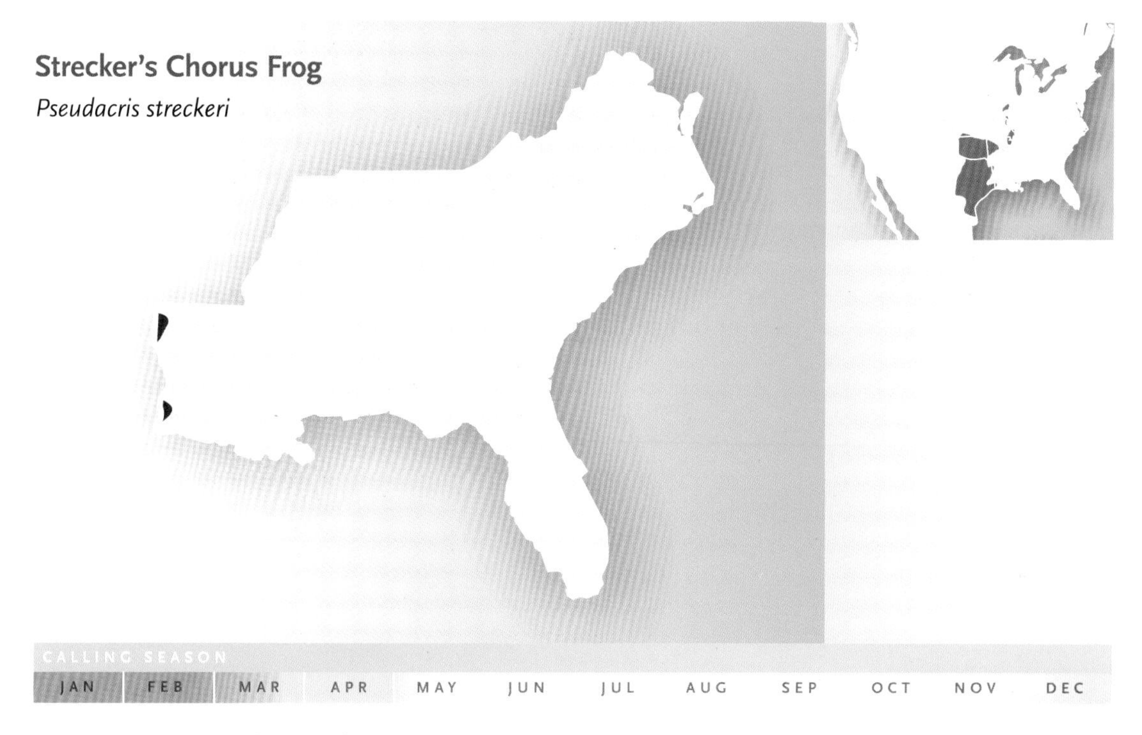

REPRODUCTION Males call from December into the spring, primarily at night but also during the daytime. They sometimes call while sitting in water but more often call from the edge of the wetland. After mating, females deposit up to 600 eggs in masses of 40 or so throughout the breeding site. The eggs and tadpoles may be freeze tolerant. The eggs hatch within a few days, and the tadpoles take about 2 months to develop, reaching a total length of about 1.5 inches before transformation. Large numbers of newly transformed froglets will often leave the wetland simultaneously and travel up to several hundred yards into the surrounding uplands.

PREDATORS AND DEFENSE Predatory fish such as sunfish and bass probably eat tadpoles and eggs. Large aquatic insect larvae such as dragonfly naiads also eat tadpoles. Snakes such as watersnakes, garter snakes, and ribbon snakes may take adults or tadpoles if the weather is warm enough for snakes to be active. Hognose snakes are known predators of adults, as are short-tailed shrews.

CONSERVATION Strecker's chorus frogs are considered a Species of Special Concern in Louisiana, primarily because of their limited range in the state. In Texas and Oklahoma, where they are relatively common and wide ranging, Strecker's chorus frogs are considered secure; however, they are considered Threatened or Endangered in most other states in which they are found.

Did you know?

All adult frogs are carnivores; that is, they eat only other animals.

Mountain chorus frogs are most active during the springtime.

How do you identify a mountain chorus frog?

SKIN
Granular

LEGS
Long and slender

FEET AND TOES
Limited webbing on hind feet; small toe pads

BODY PATTERN AND COLOR
Gray or tan

DISTINCTIVE CHARACTERS
Inward-curving stripes usually present on back; dark triangle between eyes

CALL
Raspy, fast trill repeated rapidly

SIZE
max tadpole = 1.5"
typical adult = 1"

Mountain Chorus Frog *Pseudacris brachyphona*

DESCRIPTION These small, somewhat stocky frogs are usually brown, olive, or gray. Most adults have two stripes that run the length of the back and curve inward in the middle like two reversed parentheses. These stripes are sometimes missing or broken, or they can touch to form an X on the back. A dark mask runs from the snout through the eyes but usually terminates just past the eardrums. A dark triangle is usually present between the eyes, and there are dark bands on the hind legs. These frogs generally have a light-colored upper lip, as do many chorus frogs, and yellow on the hidden parts of their thighs. The male's belly is lighter in color than the female's, and his throat is typically darker. The toe pads are fairly evident, though very small, and approach the size of those of spring peepers.

WHAT DO THE TADPOLES LOOK LIKE? The tadpole is relatively small and light brown, often mottled with light specks. The belly is lighter, and the musculature in the tail has only diffuse pigment. The tail fin is transparent. Tadpoles can reach a length of about 1.5 inches.

SIMILAR SPECIES Mountain chorus frogs are easy to confuse with other closely related species such as spring peepers and upland chorus frogs. They can be distinguished from upland chorus frogs by their enlarged toe pads and the two inward-curving stripes on the back; upland chorus frogs have three stripes that do not curve inward. Spring peepers generally have a distinct X on the back, no white on the upper lip, and more extensive webbing between the toes on the hind feet. Some people might confuse mountain chorus frogs with wood frogs, with which they share similar habitats. Wood frogs get much larger, however, and have dorsolateral ridges along each side.

DISTRIBUTION AND HABITAT Mountain chorus frogs are found throughout the Appalachian Mountains from Pennsylvania southward into northern Alabama. In the Southeast, they are found in west-central Tennessee, extreme western North Carolina, northern Georgia, and the northern half of Alabama. They barely range into extreme northeastern Mississippi and an adjacent county in southwestern Tennessee. Mountain chorus frogs are found in forested areas at elevations up to 3,500 feet in the Appalachians, and at lower elevations on the Cumberland Plateau in Tennessee. They breed in small woodland pools, bogs, woodland springs, ephemeral wetlands, furrows in plowed fields, roadside ditches, and even within pools formed in ruts on dirt roads.

Mountain chorus frog tadpoles take about 2 months to transform into adults.

BEHAVIOR AND ACTIVITY Mountain chorus frogs are active during relatively cold periods in the mountains, beginning their breeding season in February and continuing through April and even into June. They often move en masse from forests to breeding sites where

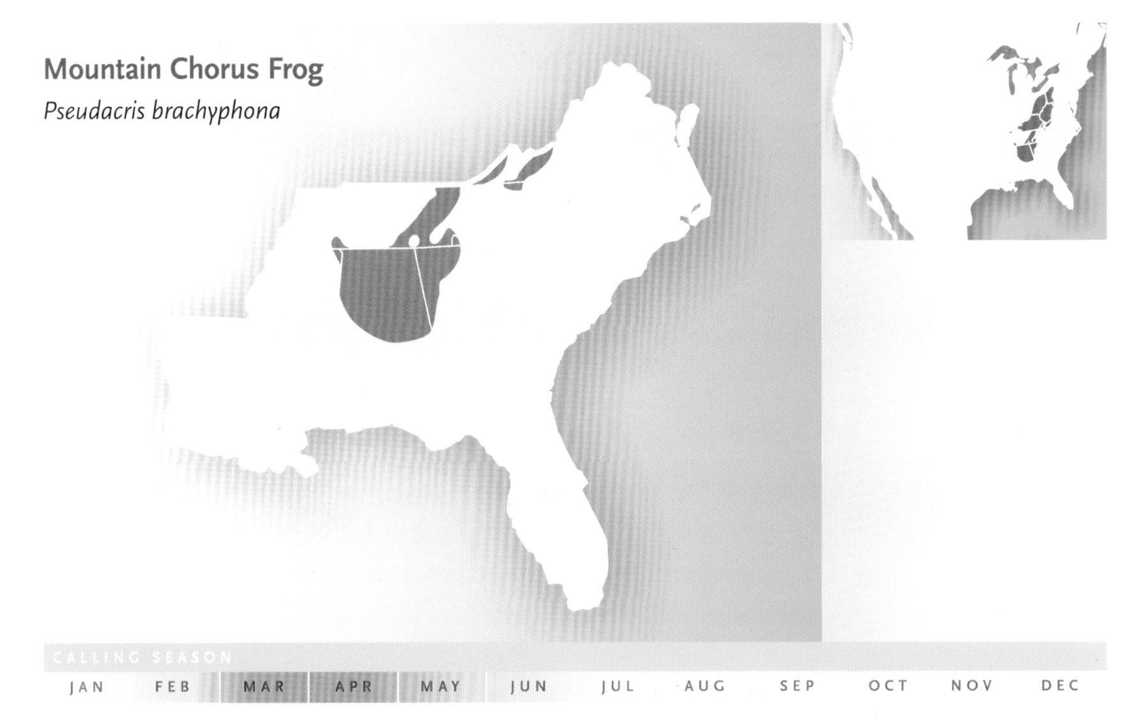

males begin calling from the edges of pools and wetlands. The males often call from the open, not concealing themselves as other chorus frogs do. They frequently call during the daytime because nighttime temperatures can drop quickly.

FOOD AND FEEDING Mountain chorus frogs feed on small, mainly terrestrial insects and other arthropods such as ants, small beetles, leafhoppers, caterpillars, spiders, and flies. They have also been known to eat earthworms and centipedes. These little frogs probably forage in upland habitats away from the pools where they breed. The tadpoles eat detritus from grass stems and other objects.

DESCRIPTION OF CALL The call is similar to that of the upland chorus frog, but the trill rate is faster and more raspy. The call is repeated much more rapidly as well, about 50–70 times per minute. Although the call also resembles that of Brimley's chorus frog, the geographic ranges of the two species do not overlap.

Very little is known about the ecology of mountain chorus frogs.

REPRODUCTION Males call from February through the spring. Females select males at the breeding site and lay eggs in small masses attached to vegetation. The masses typically contain 10–50 eggs and may be deposited on the bottom of the pool if aquatic vegetation is lacking. Females lay between 300 and 1,500 eggs per season. The eggs hatch in a week to 10 days, and the tadpoles transform within about 1–2 months, faster at warmer temperatures. The young frogs are about half an inch long at metamorphosis.

PREDATORS AND DEFENSE Bullfrogs are the only documented predators of this species, but a number of other animals probably eat them as well, including various species of birds, northern watersnakes, aquatic insects, and mammals such as raccoons and opossums. Mountain chorus frogs can jump surprisingly far for such small frogs, and that is their primary defense. They are not known to be toxic, in the adult or tadpole stage, to any predator.

CONSERVATION North Carolina considers the mountain chorus frog a Species of Special Concern because of its extremely limited distribution within the state. Deforestation, urbanization, and loss of wetland pools contribute to the decline of this species, and it has disappeared from some areas in West Virginia. For a time it was thought to have disappeared from North Carolina as well, but it was rediscovered in 2001. In other parts of the Southeast, the species appears to be fairly secure, and it is not considered a species of conservation concern anywhere but in North Carolina.

The spring peeper usually has an X-shaped marking on the back.

How do you identify a spring peeper?

SKIN
Mostly smooth

LEGS
Moderately long and slender

FEET AND TOES
Slight webbing and small toe pads

BODY PATTERN AND COLOR
Brown or tan, sometimes reddish

DISTINCTIVE CHARACTERS
Dark brown X on back

CALL
High-pitched, ascending *peeeep*

SIZE
max tadpole = 1.5"
typical adult = 1"

Spring Peeper *Pseudacris crucifer*

DESCRIPTION Spring peepers are relatively small, somewhat delicate frogs. Typical adults are tan or reddish brown in overall color when seen from above, although some may be olive or even yellowish brown. Individuals are generally lighter when they are active at night and darker when inactive or during cold weather. They normally have dark lines on the back that cross to form an X; hence the species name *crucifer*. In some individuals the X is irregular or missing altogether. A broad, dark line usually runs from the tip of the nose through the eye and along the lower side of the body. The belly is cream or whitish, sometimes yellow, and is sometimes mottled. Well-developed toe pads allow these frogs to climb very well. The toes are mostly unwebbed. Most individuals have a light lateral stripe extending behind each eye and a dark area on each side in between each eye and the snout. The vocal sac of males is generally darker than the rest of the belly.

Some spring peepers have little to no marking on their backs.

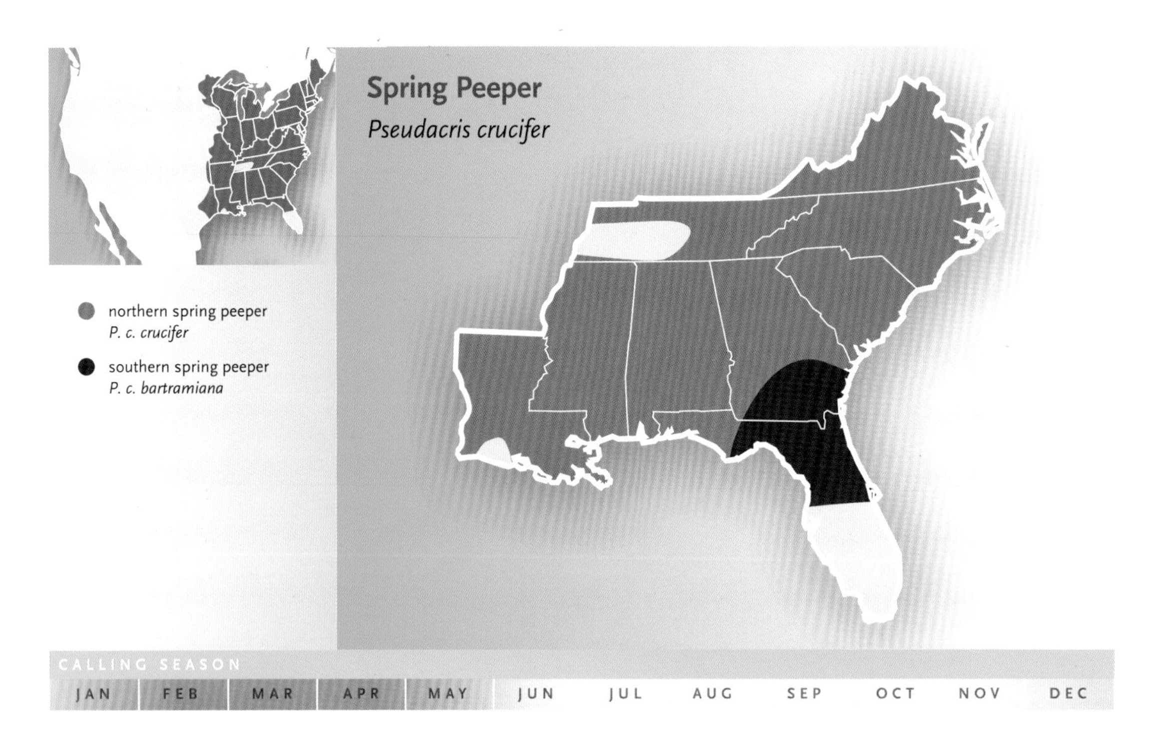

A spring peeper tadpole

The subspecies known as the southern spring peeper (*P. c. bartramiana*) occurs in southern Georgia and northern Florida and has more spotting on the belly than the northern subspecies (*P. c. crucifer*).

WHAT DO THE TADPOLES LOOK LIKE? The tadpoles are a rather nondescript olive or brown. The tail fin is moderately high and usually has faint spotting or banding but is otherwise transparent. The tail musculature is light colored. The tadpoles reach a maximum size of about 1.5 inches before metamorphosis.

SIMILAR SPECIES Spring peepers may be confused with other chorus frogs (genus *Pseudacris*), but the well-developed toe pads distinguish them immediately. Their call is similar to that of the ornate chorus frog (*P. ornata*), but each note of the spring peeper's call rises in pitch while each note of ornate chorus frogs remains at the same pitch. They might also be confused with smaller treefrogs such as pine woods treefrogs or squirrel treefrogs, but neither of these has an X on the back.

DISTRIBUTION AND HABITAT Spring peepers range across most of the eastern United States and Canada and are present throughout the Southeast except for the lower two-thirds of the Florida peninsula. They are generally found in forested habitats throughout their range and breed in a variety of temporary and permanent wetlands. They appear to be most abundant in areas with many shallow, fish-free wetlands in which their larvae are

Spring peepers are often heard calling during late winter and early spring.

Spring peeper eggs are laid in small groups attached to aquatic vegetation.

relatively safe. Breeding habitats include marshes, Carolina bays, roadside ditches, and farm ponds. They tend to avoid floodplain forests. They tolerate moderate levels of urbanization and often breed in ponds and slow-moving streams in suburban neighborhoods.

BEHAVIOR AND ACTIVITY This species is active year-round in the Southeast but is most commonly encountered in winter and early spring. Spring peepers are fairly arboreal, but they do not ascend as high into trees as typical southeastern treefrogs (genus *Hyla*) do. During the coldest parts of the winter, they retreat to shallow burrows or hide under bark. They also use such hiding places during dry, hot periods to avoid drying out. During the winter they can withstand freezing of much of their body water because they use glucose (i.e., sugar) as an antifreeze to help protect their cells from freeze damage. These little frogs are usually among the first to start breeding when environmental temperatures remain above freezing, and they can often be heard calling in early January or even December in the Southeast. At higher latitudes or elevations, they actively begin calling in early spring, earning them the name spring peepers. The tadpoles sometimes form dense schools that presumably offer some protection from predators.

FOOD AND FEEDING Spring peepers feed on a variety of small insects and other slow-moving, crawling arthropods, including beetles, ants, mites, and spiders. They apparently do not feed on aquatic prey. Seeds found in the stomachs of some individuals were probably mistaken for prey or consumed incidentally. Smaller individuals tend to prey on smaller invertebrates. Spring peepers may feed during the day and at night.

DESCRIPTION OF CALL Spring peepers get their name from the sound of their call, which is best described as a repeated, high-pitched *peeeep* whose pitch rises toward the end of the note. The sound made by individuals that are just beginning to call is sometimes more like a trill that resembles the calls of several other species of chorus frogs. When approached by another male, a calling male may give a rapid series of harsher loud trills that resemble the advertisement calls of upland chorus frogs, presumably to discourage the intruder from advancing farther. The advertisement call is very loud for such a small frog, and dense choruses can be deafening. Choruses may stop when disturbed by a human intruder or other animal, but the frogs soon begin to call again if the disturbance subsides.

The southern spring peeper is found in southern Georgia and northern Florida.

REPRODUCTION Females tend to select males that call on low vegetation 1–4 feet above the water's surface. The female lays eggs either singly or in small clumps attached to aquatic vegetation. A single female may lay up to 1,000 eggs, but 700 is a typical number. Eggs hatch in a few days to 2 weeks, and the tadpoles metamorphose after about 3 months. Froglets emerge from the water at a length of about half an inch.

PREDATORS AND DEFENSE Fish, semiaquatic snakes, and numerous predatory insects eat spring peepers. Larger frogs and various species of birds probably eat them as well. Their coloration provides camouflage for these small, relatively inconspicuous frogs, which are not known to be particularly toxic. When disturbed, individuals can jump considerable distances.

CONSERVATION Spring peepers are extremely abundant throughout most of their range, even though they may remain hidden during the warmer months. Because of their abundance and the lack of evidence of any widespread declines in the Southeast, they are not considered a species of conservation concern in any southeastern state. They are protected in New Jersey and Kansas, where populations are either small or have apparently declined. Extensive disturbance of wetlands and associated forested habitat surely has resulted in local declines of this species in various places throughout the Southeast.

COMMENTS Because of its arboreal habits and well-developed toe pads, this species was once considered more closely related to treefrogs of the genus *Hyla* and was known as *Hyla crucifer*. Studies during the 1980s showed that spring peepers are actually more closely related to chorus frogs, and the species was placed in that genus.

Green treefrogs are found throughout the Southeast.

Green Treefrog

Hyla cinerea

DESCRIPTION Green treefrogs are medium to large, somewhat slender frogs characterized by a solid green back and a well-defined ivory to yellow stripe running from the upper lip along each side nearly to the groin. In some populations the stripe may be reduced or absent. The background color may vary from brownish green to bluish gray when individuals are concealed or inactive to bright lime green in active individuals. Many green treefrogs have small, well-defined orange spots on the back that may be inconspicuous in some individuals. The underside is white or cream. Males usually have a darker throat than females.

WHAT DO THE TADPOLES LOOK LIKE? The tadpoles have high, spotted or reticulated tail fins and a light stripe from the eye to the tip of the snout. The tail is relatively long.

SIMILAR SPECIES Green treefrogs can be distinguished from squirrel and barking treefrogs by the solid green back and well-defined light stripe along each side. In

Green treefrogs often have small but distinct yellow, gold, or orange spots on their backs.

How do you identify a green treefrog?

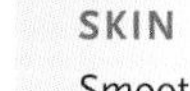

SKIN
Smooth

LEGS
Long and slender

FEET AND TOES
Moderate webbing on hind feet; large toe pads

BODY PATTERN AND COLOR
Bright green back, often with tiny orange dots

DISTINCTIVE CHARACTERS
Well-defined light stripe along each side

CALL
short, loud *quenk* that may sound like a cowbell

SIZE
max tadpole = 2"
typical adult = 2"

The call of the green treefrog may resemble cow-bells from a distance.

Did you know?

*The crab-eating frog (*Rana = Fejervarya cancrivora*) of Singapore is one of the few amphibians in the world that can tolerate seawater.*

addition, green treefrogs are more slender and have smoother skin than barking treefrogs and are larger than squirrel treefrogs. No other south-eastern treefrog has small, bright orange or gold spots.

DISTRIBUTION AND HABITAT Green treefrogs range from the Chesapeake Bay southward to the southern tip of Florida and westward into eastern Texas, southeastern Oklahoma, much of Arkansas, and up the Mississippi drainage to southern Illinois, southeastern Missouri, and eastern Tennessee. In the Southeast, they are found primarily in the Coastal Plain, although in some areas (e.g., North Carolina) they appear to be expanding their range well into the Piedmont. Green treefrogs are present in and around a variety of aquatic habitats including swamps, isolated wetlands, ponds, lakes, and rivers. They tend to be more willing to breed in waters inhabited by fish than are other treefrogs. Green treefrogs are often the only species of treefrog found on barrier islands, where they have been reported to exhibit some degree of saltwater tolerance, at least for short periods.

BEHAVIOR AND ACTIVITY Seasonally, green treefrogs are generally active from mid-spring until late summer or early fall. They usually remain hidden during the daytime, perched high in trees or under bark or other vertical surfaces such as loose boards on buildings, although they can sometimes be found in the open, sitting motionless on a tree, cattail frond, or other

Green Treefrog

Hyla cinerea

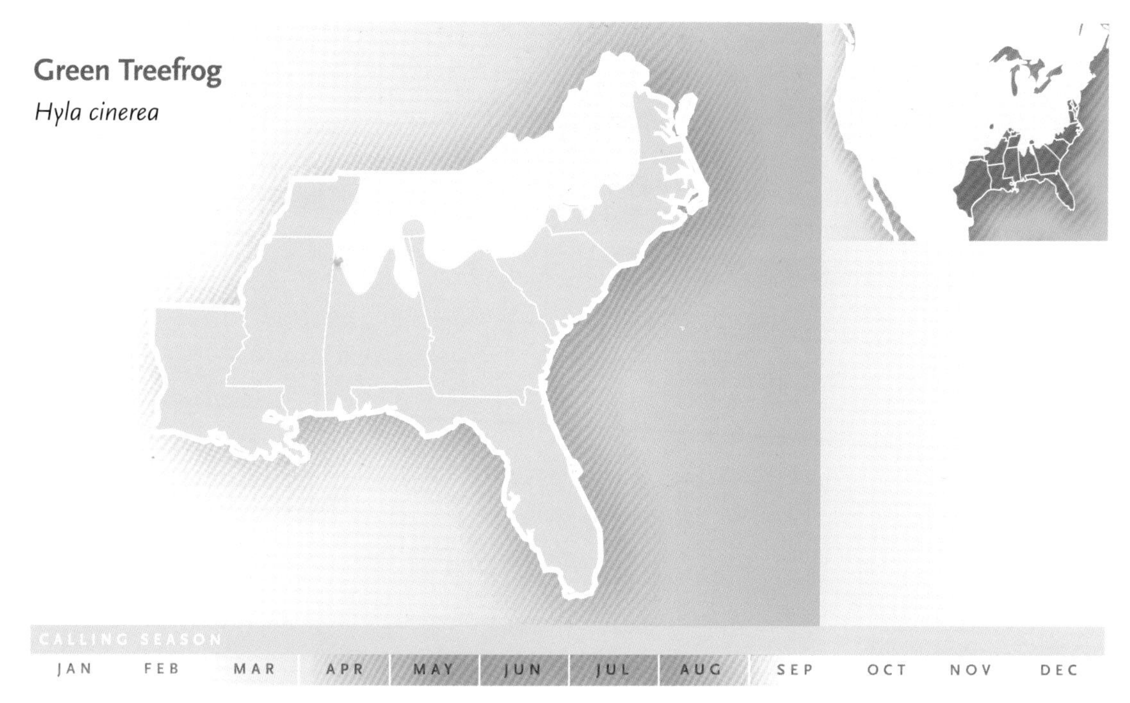

This green treefrog lacks yellow pigment across much of its body, resulting in a bluish coloration.

vertical surface with their legs tucked under their body, even basking in complete sunlight. They become active at dusk, often descending from treetops to feed or engage in breeding activity, and are often active at night around houses, where they hunt for insects at lights.

FOOD AND FEEDING Green treefrogs feed almost exclusively on insects and other arthropods; about 90 percent of the prey items recorded are beetles, moths, and orthopterans such as katydids and grasshoppers. The diet also includes spiders, dragonflies, various flies, and even snails. Green treefrogs generally hunt from a perch, waiting for unsuspecting prey to approach and then jumping and seizing the victim in their mouth.

DESCRIPTION OF CALL The advertisement call is a loud, short, nasal, duck-like *quenk* that from a distance reminds some people of cowbells. When a male is just getting started, the call may be quieter and atypical. A male approached too closely by another male will produce a rapid succession of short bursts and sometimes trills apparently meant to drive the intruder away. Green treefrogs often call in very large choruses that sound like a constant roar from far away and may even be unrecognizable as frogs by people unfamiliar with the species. During the height of their breeding activity, males are usually determined to call. When disturbed by an observer, a male will often jump to a new location and continue to call without missing a beat. Calling activity in large groups occurs in waves. As part of the group reaches a peak in calling activity, nearby groups begin to call; meanwhile, the first group eventually ceases to call, and this "wave" of calling activity passes through the entire chorus.

REPRODUCTION Breeding activity generally begins in spring, earlier in southern regions and later in the more northern parts of the Southeast, and generally continues through the summer. Adults migrate from surrounding forests to breeding sites where the females select mates and initiate amplexus. The eggs are attached to floating vegetation in clusters of about 500 to more than 1,000 eggs. A single female may lay multiple clutches of eggs during a breeding season, not always in one area. The tadpoles grow to about 2.5 inches and transform into the adult form within 50–60 days. Green treefrogs interbreed with barking treefrogs in some areas, and research has shown that habitat alteration such as clearing of vegetation can facilitate hybridization.

Green treefrogs are sometimes eaten by snakes (like this common garter snake) that prowl the edges of wetlands at night.

PREDATORS AND DEFENSE Green treefrogs are susceptible to predation by many small to medium-sized vertebrates such as watersnakes and juvenile rat snakes. Large fish, turtles, small alligators, and aquatic insect larvae probably eat the tadpoles. Adults rely on their leaping ability to evade most predators; their skin toxins are apparently not very potent.

CONSERVATION Green treefrog populations appear to be fairly secure throughout most of their range, although destruction or contamination of wetlands and other habitats will likely eliminate or reduce the size of some populations. In some areas they appear to be expanding their range. For example, green treefrogs were not recorded near Davidson, North Carolina, until 2001, but by 2005 numerous new locality records had been reported. This species probably hitchhikes on horticultural plants, gardening supplies, and lawn equipment.

Some people consider the Pine Barrens treefrog to be the most beautiful frog in North America.

Pine Barrens Treefrog *Hyla andersonii*

DESCRIPTION Many people consider the Pine Barrens treefrog the most beautiful frog in the United States. The back is typically green to dark green. A dark stripe bordered by white extends from the nostril through the eye, widening as it continues on each side. The dark stripe is usually brownish but may have a purplish tinge. The underside is generally white with considerable orange spotting in the groin and concealed parts of the legs. Orange spotting may also be present around the shoulders and on the feet. Males have a darker throat than females, especially during the breeding season.

WHAT DO THE TADPOLES LOOK LIKE? Pine Barrens treefrog tadpoles are usually brown or golden. The tail musculature is spotted with several large, dark blotches. The tail fin is rather dark. The tadpoles reach a maximum size of about 1.5 inches.

SIMILAR SPECIES Pine Barrens treefrogs are most easily confused with green treefrogs, barking treefrogs, and squirrel treefrogs, but they can be distinguished from all three by the wide, dark stripe that runs down each side. The call can be confused with that of the green treefrog, but it is higher pitched, and usually the notes are repeated more rapidly.

How do you identify a Pine Barrens treefrog?

SKIN
Smooth

LEGS
Relatively long and slender

FEET AND TOES
Moderate webbing on hind feet; large toe pads

BODY PATTERN AND COLOR
Green to bright green; dark brown on sides with stripe through eye

DISTINCTIVE CHARACTERS
Well-defined dark stripe running from nose through eye and down each side

CALL
Short, nasal *quenk* repeated once or twice per second

SIZE
max tadpole = 1.5"
typical adult = 1.5"

DISTRIBUTION AND HABITAT Pine Barrens treefrogs are found in three separate and widely disconnected areas of the Southeast: the western panhandle of Florida and adjacent southern Alabama, the region encompassing the northern sandhills of South Carolina, and the sandhills and parts of the Coastal Plain of North Carolina. The species is also present in the Pine Barrens of New Jersey, the area from which it gets its common name. A record from Georgia has never been verified. Within the Carolinas, Pine Barrens treefrogs breed primarily in sandhills areas with shrub-herb bogs along small black-water tributaries. A key feature of the aquatic breeding site is acidic water, generally with a pH less than 4.5. The wetland habitats are usually thick with sphagnum or other aquatic vegetation such as sedges, rushes, bladderwort, pipeworts, sundews, pitcher plants, club moss, and filamentous algae. Breeding habitats in Florida and southern Alabama are typically seepage wetlands in sandhills and clayhills, usually those associated with hillside seepages, and sometimes seeps associated with disturbance that disrupts the ability of hardwood trees to become established, such as in power line rights-of-way. Historically, natural fires were important in preventing hardwood shrubs from dominating the herbaceous portion of seepage bogs.

BEHAVIOR AND ACTIVITY Pine Barrens treefrogs are active from March until November in the Southeast, although their breeding season typically

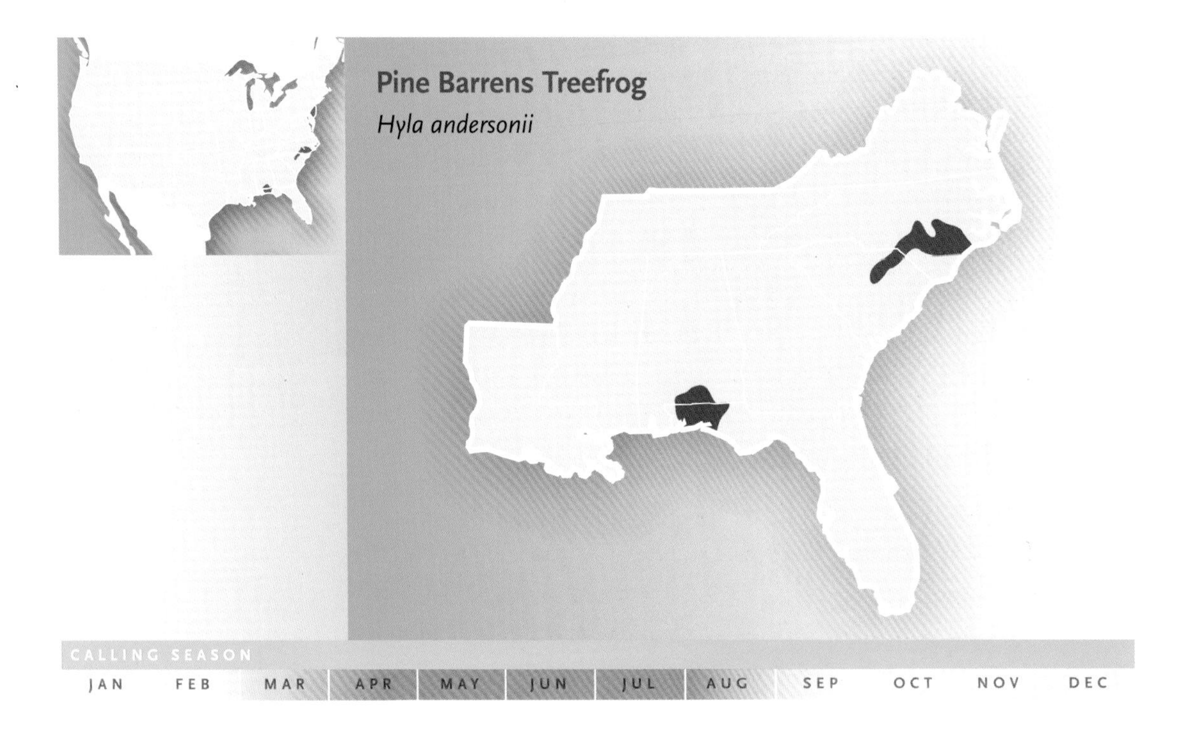

Some Pine Barrens tree-frogs have orange spotting around the shoulders and on the feet.

begins in April and continues into midsummer. When not involved in breeding, adults apparently spend most of their time in habitat within or close to their breeding site, although some adults have been known to disperse more than 100 yards from breeding ponds within a month after breeding. They retreat to various hiding places in wetland shrubs during the day and apparently spend the winter hiding under loose bark or in other inconspicuous places.

A male Pine Barrens treefrog calls from a perch above a small wetland.

FOOD AND FEEDING Like other treefrogs, Pine Barrens treefrogs eat primarily insects, which they likely capture by ambush. Animals from New Jersey prey on grasshoppers, beetles, ants, and flies; those in the Southeast presumably have a similar diet, along with moths and any other insect they happen to encounter. The tadpoles are herbivorous and feed on algae.

DESCRIPTION OF CALL The call is a nasal *quenk-quenk-quenk* repeated rapidly, often more than once per second; the repetition rate is slower at lower temperatures. Calling bouts are somewhat infrequent and occur at irregular intervals. The call is fairly loud, but not as loud as that of the green treefrog, which it somewhat resembles. Some people think it sounds like the horns found on children's bicycles.

REPRODUCTION Breeding activity typically begins in mid-to-late April and continues through midsummer, primarily during rainy periods. Females lay between 500 and 1,000 eggs, generally depositing them one at a time and attaching them to vegetation or to debris on the bottom of the wetland. Eggs hatch within a few days, and the tadpoles metamorphose after about 2 months.

PREDATORS AND DEFENSE Northern watersnakes and ribbon snakes are documented predators; larger frogs, other species of snakes, small to medium-sized mammals, and predatory birds probably eat these treefrogs as well. The tadpoles likely fall prey to large aquatic insects and possibly salamanders, red-fin pickerel, and turtles such as spotted turtles that inhabit similar habitats.

Pine barrens treefrogs are rare, and areas where they occur are widely separated geographically.

CONSERVATION Like most animal species in the Southeast, populations of Pine Barrens treefrogs have suffered because of agriculture, urban and industrial development, and succession of their breeding habitats due to disruption of the natural fire cycles. Because they have very specific habitat requirements and occur in isolated and often small populations, Pine Barren treefrogs may be particularly vulnerable to local extinction. Conservation depends on preservation and management of breeding habitats by periodic prescribed burns. Threats include any alteration of breeding habitats and the nearby uplands. Pine Barrens treefrogs are listed as Significantly Rare in North Carolina and as Threatened in South Carolina. The federal government once listed Pine Barrens treefrogs as Endangered in Florida, but the species was delisted when more information about its distribution became available. The Pine Barrens treefrog is now listed as Rare by the state of Florida, where more than 135 localities have been documented, and is considered Threatened by the state of Alabama.

This barking treefrog's upper lip and side show the white line typical of the species.

How do you identify a barking treefrog?

SKIN
Granular

LEGS
Moderately stout but long

FEET AND TOES
Moderate webbing on hind feet; large toe pads

BODY PATTERN AND COLOR
Gray to bright green; spots usually covering back; light stripe along each side

DISTINCTIVE CHARACTERS
Large size and dark spots on back; granular texture to skin

CALL
Resembles a barking dog

SIZE
max tadpole = 2.75"
typical adult = 2.25"

Barking Treefrog *Hyla gratiosa*

DESCRIPTION Our largest native treefrog, barking treefrogs are more robust than most other treefrog species and have a shorter, squatter body. Their skin is somewhat rough or granular. The back ranges from gray to bright green, even in the same individual, and is covered with small to medium-sized dark brown spots. Under certain conditions, individuals may become bright green and the dark spots may fade or even disappear completely. The upper lip usually has a white line, which often extends along the side of the body as an irregular white line above a brown area. The hind legs have dark bars. The belly is white or cream, sometimes with a tinge of pink, and the groin is yellowish. Males have a darker throat than females.

WHAT DO THE TADPOLES LOOK LIKE? Barking treefrog tadpoles are larger than other treefrog tadpoles. The back is olive to olive brown, the belly is yellowish, and the tail fin is very tall. Small individuals may have four dark stripes on the back, and larger ones often have a dark saddle about halfway down the tail.

A tadpole of the barking treefrog

SIMILAR SPECIES Barking treefrogs might sometimes be confused with green treefrogs, but the latter are slimmer in build; have very smooth skin (i.e., not granular); and nearly always have a well-defined, smooth-edged white stripe running down the side of the body. Southern leopard frogs and pickerel frogs are also spotted, but they get larger than barking treefrogs and lack toe pads. Young barking treefrogs can sometimes be confused with squirrel treefrogs, but their more robust appearance and more granular skin will distinguish them from that species.

Barking treefrogs call while floating in the water. Their call resembles the bark of a dog.

DISTRIBUTION AND HABITAT Barking treefrogs are found in the Coastal Plain from North Carolina south through all but south-central Florida and east to southeastern Louisiana. Isolated populations in Kentucky, Tennessee, New Jersey, Virginia, and Delaware suggest that this species was once more widespread in the eastern United States. Barking treefrogs inhabit forested areas and usually occur near or within shallow wetlands. They can be abundant in Carolina bays and other ephemeral wetlands that do not contain fish, but in some situations they reproduce in wetlands where fish are present. Adults can also be found around the margins of river floodplains, where they breed in temporary wetlands.

The spots of the barking treefrog often fade at night when this species is more active.

Barking Treefrog

Hyla gratiosa

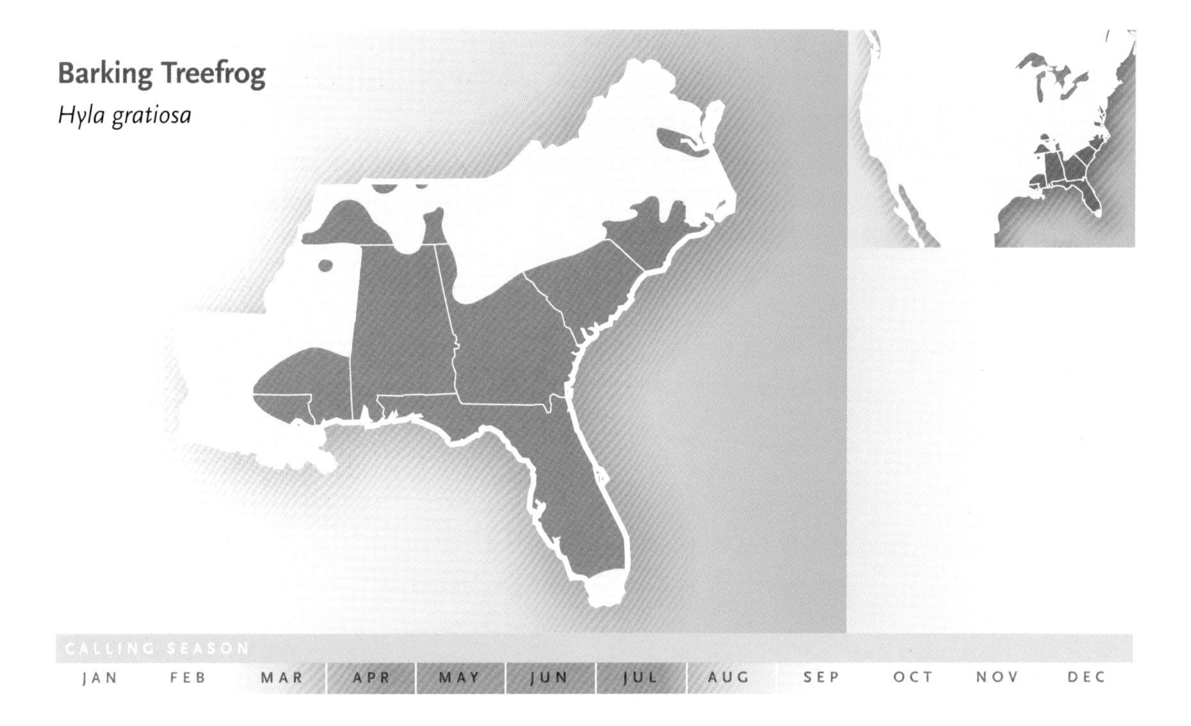

BEHAVIOR AND ACTIVITY Barking treefrogs spend less time in trees than other treefrogs do. They are generally active at night. During the day they typically seek cover in relatively low trees or shrubs but will sometimes bask in the sun. They may also burrow into damp soil or under logs near wetlands or hide under loose bark on upright pine stumps. During winter they burrow into soil or take shelter in burrows of other animals such as gopher tortoises or small mammals. They are often docile when handled and will sit on a person's hand or finger without attempting to flee. Calling males are very sensitive to disturbance and will usually cease calling immediately and dive when approached. When disturbed, barking treefrogs often change color rapidly from green to gray or brown.

FOOD AND FEEDING Like most frogs, barking treefrogs eat any animal they can swallow, but their diet consists mainly of insects such as moths and beetles. They hunt primarily by ambush, waiting for an unwary insect to venture close enough to be eaten. They may use their forelimbs when handling particularly feisty prey.

DESCRIPTION OF CALL The advertisement call is lower pitched than other treefrogs' calls and from far away sounds remarkably like a barking dog. The calls run together into an irregular roar in very large choruses, and standing in a wetland amid hundreds of calling barking treefrogs can be a powerful, and nearly deafening, experience. Males usually call while floating

Did you know?

The toe pads of treefrogs act like little suction cups and allow these frogs to climb smooth surfaces such as glass windows.

on the surface of the water at night, often holding onto some sort of floating vegetation with their front legs; other treefrogs generally call near water or in emergent vegetation, but not while floating. They sometimes call from high in the treetops at dusk or even during the middle of the day.

REPRODUCTION Adult males begin calling in March or April, and breeding may continue until late summer. Adults of both sexes migrate from their winter retreat sites to nearby wetlands, sometimes using multiple breeding sites within their lifetime. Barking treefrogs generally begin reproduction later at higher latitudes, and in the southern part of their range (e.g., southern Florida) may call into early fall. Each female selects a mate, presumably based on the quality of his call, and allows him to engage in amplexus. Eggs are deposited singly or in clusters that are attached to submerged vegetation or rest on the bottom of the wetland. After mating, the male and female generally leave the wetland and spend the rest of the year in the adjacent forest.

PREDATORS AND DEFENSE Predators of adults include watersnakes of various species, cottonmouths, and other snakes that inhabit wetland habitats. Southern hognose snakes have been recorded feeding on barking treefrogs that they may have uncovered while burrowing. Small to medium-sized mammals and birds are likely predators as well. Large aquatic invertebrates, turtles, and watersnakes probably eat the tadpoles. When threatened, adults inflate their body with air. Their skin may be somewhat toxic to the mucous membranes of many predators. The saddle marking on larger tadpoles may distract predators and allow the tadpole to escape.

Barking treefrogs have a shorter, squatter body than most other treefrog species.

CONSERVATION Barking treefrogs can be common in areas with ephemeral wetlands and forested habitats. When wetlands and forests are destroyed or altered, though, this species may be one of the first treefrogs to disappear. Barking treefrogs are listed as Endangered in Delaware and Threatened in Virginia, and are protected in Tennessee and Maryland, primarily because they occur in small, isolated populations in these states. Development and highways can isolate uplands populations from their breeding sites, making them more vulnerable to extinction. Conservation of barking treefrogs should consist of preserving complexes of ephemeral wetlands and the surrounding terrestrial habitats. These large, attractive treefrogs, with their tolerance for being handled, are often collected for the pet trade, which could result in local population declines.

Distinguishing a Cope's gray treefrog from a common gray treefrog is difficult unless you hear its call.

Common Gray Treefrog *Hyla versicolor*
Cope's Gray Treefrog *Hyla chrysoscelis*

DESCRIPTION Because these two species are essentially impossible to distinguish without examining them under a microscope or hearing their call, we consider them together in a single account. Both are somewhat stocky treefrogs that are predominantly gray, generally with a large, irregular blotch on the back. The body color of an individual frog can change dramatically from dark to light gray or even green depending on temperature and activity level. A light spot is present under each eye, and the skin is covered with small warts or bumps that give it a rough texture. The skin of inactive individuals sometimes takes on a smoother, glossy sheen. Dark bars are present on the hind limbs, and the groin and concealed surfaces of the thighs are bright orange-yellow. Females grow larger than males.

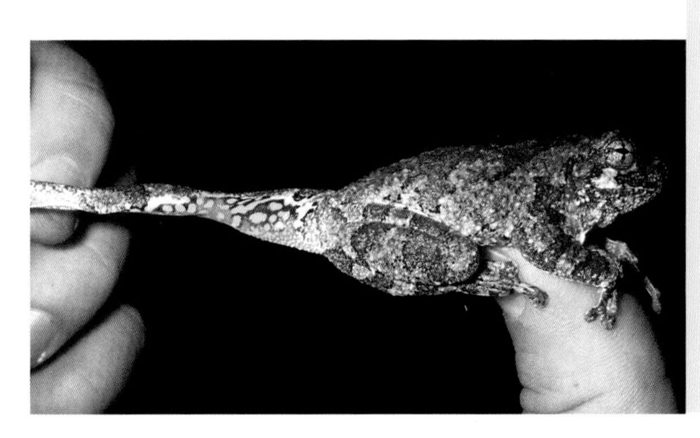

The bright orange coloration on the hidden parts of the hind legs of gray treefrogs is thought to startle potential predators.

How do you identify a gray treefrog?

SKIN
Granular

LEGS
Moderately long

FEET AND TOES
Moderate webbing on hind feet; large toe pads

BODY PATTERN AND COLOR
Mottled gray or brown, usually with a large, irregular blotch on the back

DISTINCTIVE CHARACTERS
Inner thighs and groin yellowish orange; light spot under each eye

CALL
Musical trill; slower in common gray treefrogs; faster and harsher in Cope's gray treefrogs

SIZE
max tadpole = 2"
typical adult = 1.75"

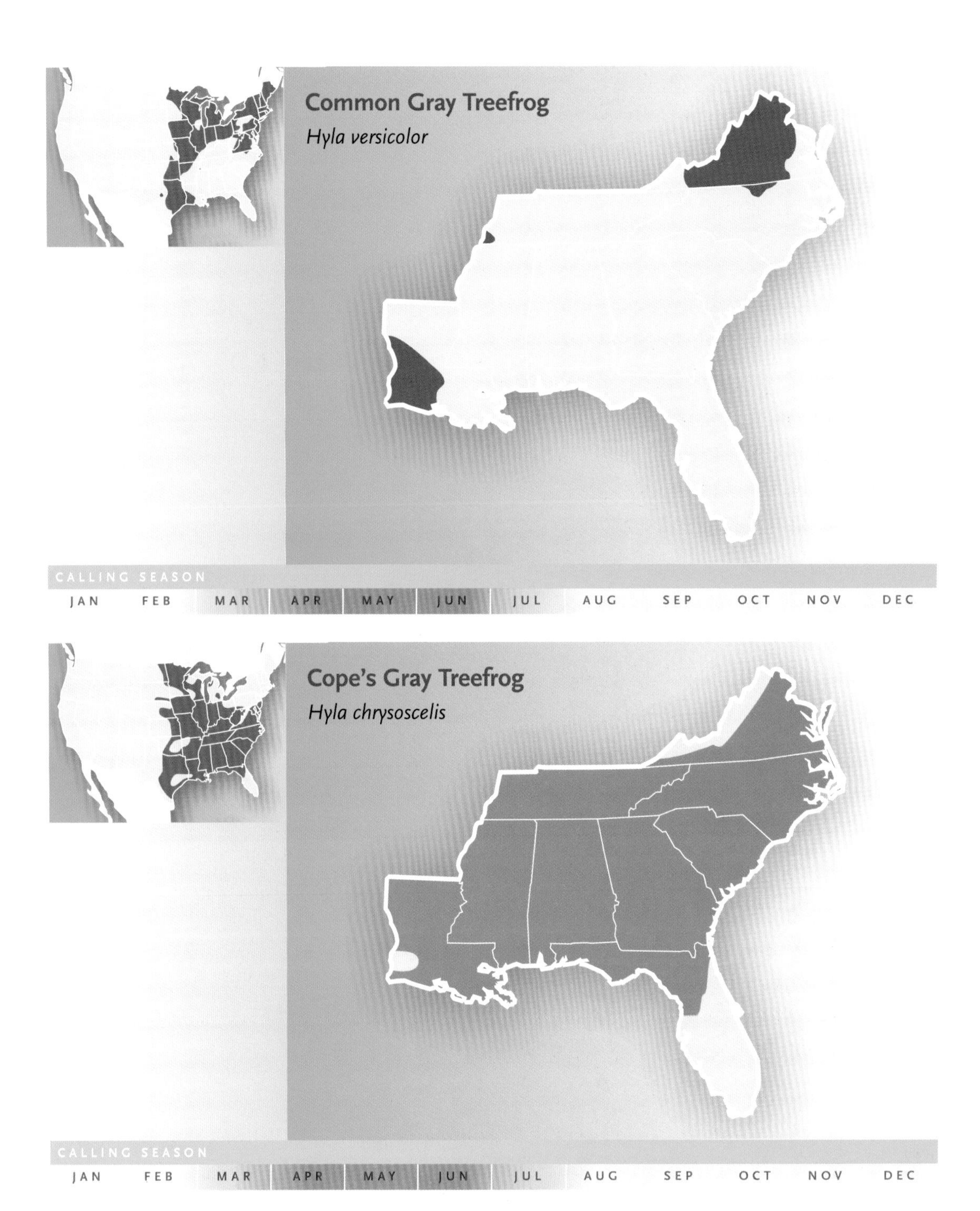
Common Gray Treefrog
Hyla versicolor
CALLING SEASON
JAN
FEB
MAR
APR
MAY
JUN
JUL
AUG
SEP
OCT
NOV
DEC
Cope's Gray Treefrog
Hyla chrysoscelis
CALLING SEASON
JAN
FEB
MAR
APR
MAY
JUN
JUL
AUG
SEP
OCT
NOV
DEC

A Cope's gray treefrog tadpole

Cope's gray treefrogs can be distinguished from common gray treefrogs on the basis of their chromosome number. Cope's gray treefrogs have 24 chromosomes—two copies of each one—and are thus diploid (2N). Common gray treefrogs have twice as many chromosomes; that is, they have four copies of each chromosome and are tetraploid (4N). The two species can be distinguished by examining cell size under a microscope; in general, the cells of common gray treefrogs are larger than comparable cells of Cope's gray treefrogs.

WHAT DO THE TADPOLES LOOK LIKE? Tadpoles of both species are usually somewhat gray or brown and have a high, mostly translucent tail fin spotted with black. In some regions, the tail fin of larger tadpoles is often reddish or orange; the color may be related to the presence of particular predators. The tadpoles reach a length of nearly 2 inches before transforming into froglets.

SIMILAR SPECIES Although they are difficult to distinguish from one another, the two gray treefrog species are relatively easy to distinguish from other treefrogs. Bird-voiced treefrogs are generally more delicate, have a more prominent rectangular spot under each eye, and have greenish groin and thigh regions. Pine woods treefrogs are smaller, more slender, and have dark posterior thighs with yellow or white spots. Barking treefrogs grow considerably larger and have a greenish back with many small, isolated dark spots.

DISTRIBUTION AND HABITAT Collectively, the two species of gray treefrogs range across almost the entire eastern half of the United States. One or both can be found throughout the Southeast except for peninsular Florida. Because they are so difficult to tell apart, the actual distribution of each is hard to determine exactly by casual observation. They occur together in some places and singly in others. In the Southeast, the common gray treefrog is known from southwestern Louisiana, the western half of Virginia, north-central North Carolina, and extreme southwestern Tennessee. Cope's

A Cope's gray treefrog

A green color variant of the common gray treefrog

gray treefrog is found throughout most of the Southeast, being absent only from peninsular Florida, a small area in southwestern Louisiana, and most of the mountains of Virginia. Gray treefrogs live primarily in forests, but often move into open areas such as pastures and marshes to breed. Breeding habitat includes ephemeral wetlands, roadside ditches, and the edges of ponds. Cope's gray treefrogs seem to be better at colonizing urban areas and can often be found in birdhouses, around swimming pools, and in potted plants. In some places Cope's gray treefrogs appear to prefer lower elevations while common gray treefrogs prefer higher elevations.

BEHAVIOR AND ACTIVITY When not breeding, these treefrogs usually remain high in the treetops, vocalizing from time to time, although they are sometimes found nearer to the ground. Gray treefrogs often take refuge under or within objects that offer cover, where they can remain hidden and protected from dehydration. In eastern Texas, where both species occur, the common gray treefrog tends to call from lower perches or from the ground while Cope's gray treefrog calls more often from bushes and vegetation within or adjacent to the breeding site. During the winter, both species take refuge under bark or possibly within shallow burrows. Common gray treefrogs can withstand freezing of much of their body water for several days. They produce glycerol and/or maintain high levels of glucose in their blood that help to protect their cells from freeze damage.

FOOD AND FEEDING Gray treefrogs of both species feed primarily on insects and can often be found around lighted windows and porches at night waiting in ambush for their prey, which includes caterpillars, beetles, roaches, and crickets, and probably moths as well. There is some evidence that Cope's gray treefrogs eat more arboreal insects, possibly due to their tree-climbing habits, than do common gray treefrogs. Tadpoles feed on algae and other detritus that they scrape from subsurface rocks, tree limbs, and aquatic vegetation.

DESCRIPTION OF CALL Listening to the call is perhaps the easiest way to distinguish the two species of gray treefrogs from each other. The common gray treefrog has a loud, somewhat musical trill with approximately 16–35 notes per second. The trill rate of Cope's gray treefrog is considerably faster (30–65 notes per second) and harsher; it has been likened to that of a red-bellied woodpecker. Trill rates of both species are higher at higher temperatures, but if both species are calling at the same temperature at the same locality, the differences are very apparent. When approached by another male, a gray treefrog will give a call that sounds like a hen turkey calling to her mates—*chow, chow, chow, chow*—repeated about 6–10 times within 2–3 seconds.

REPRODUCTION Gray treefrogs usually begin breeding in late March or April in the Southeast. Breeding continues throughout the summer, peaking in late April–mid-June. Each female lays up to 2,000 eggs in small clutches of 10–50 eggs each. Tadpoles are about 2 inches long at metamorphosis, and the larval period is about 2 months or shorter in warmer, more southern localities.

Gray treefrogs usually begin breeding in late March or April in the Southeast.

PREDATORS AND DEFENSE Adults probably fall prey to several species of snakes, including juvenile rat snakes and watersnakes, in addition to raccoons, skunks, and opossums. Dragonfly larvae, large aquatic beetles, fish, aquatic snakes, wading birds, turtles, aquatic salamanders, and large salamander larvae eat the tadpoles. Some studies have shown that gray treefrog tadpoles use chemical cues to detect the presence of fish. The skin of adults produces a mucus that can be an irritant to some predators. These secretions can be extremely painful if they contact the eyes or nasal membranes of humans. Some scientists believe the bright yellow "flash" color on the groin and thighs of gray treefrogs startles potential bird predators, giving the frog a chance to escape.

CONSERVATION Gray treefrogs of both species are generally abundant in most localities within the Southeast where they occur, and they appear to tolerate moderate human disturbance. Confusion between the two species makes documentation of exact geographic ranges difficult, though, which in turn complicates efforts to determine their exact status. Cope's gray treefrogs are listed as Endangered in New Jersey.

Bird-voiced treefrogs get their name from their melodious call.

How do you identify a bird-voiced treefrog?

SKIN
Smooth, but with small bumps

LEGS
Moderately long

FEET AND TOES
Moderate webbing on hind feet; toe pads on each toe

BODY PATTERN AND COLOR
Mottled light to dark gray

DISTINCTIVE CHARACTERS
White rectangular spot beneath each eye; greenish coloration on groin and undersides of thighs

CALL
High-pitched whistle, repeated several times

SIZE
max tadpole = 1.5"
typical adult = 1.5"

Bird-voiced Treefrog *Hyla avivoca*

DESCRIPTION Bird-voiced treefrogs are generally various shades of mottled gray, often with a pattern that resembles the gray or greenish lichens on an oak tree. The overall coloration varies with activity and temperature from dark gray to light gray to green. A single large, irregular dark blotch is frequently present on the back. The legs often have alternating dark and light gray bars, and the belly is white. A rectangular white spot is almost always present on the upper lip below each eye. The groin and inner surface of the hind legs are green or greenish yellow, but this distinctive coloration is usually not visible when the frog is sitting still. The skin is relatively smooth with small bumps.

WHAT DO THE TADPOLES LOOK LIKE? The tadpole is mostly black with several coppery bands on top of the tail musculature. A coppery blotch is usually visible on top of the head, and the translucent tail fin may have black mottling. The belly is black. The eyes stick out from the head when viewed from above. The tadpoles are less than half an inch long at hatching and grow to 1.5 inches before transforming into froglets.

SIMILAR SPECIES Bird-voiced treefrogs are most easily confused with common and Cope's gray treefrogs, which are somewhat larger. The greenish groin and inner thigh, as opposed to the yellow-orange found in both gray

Bird-voiced Treefrog

Hyla avivoca

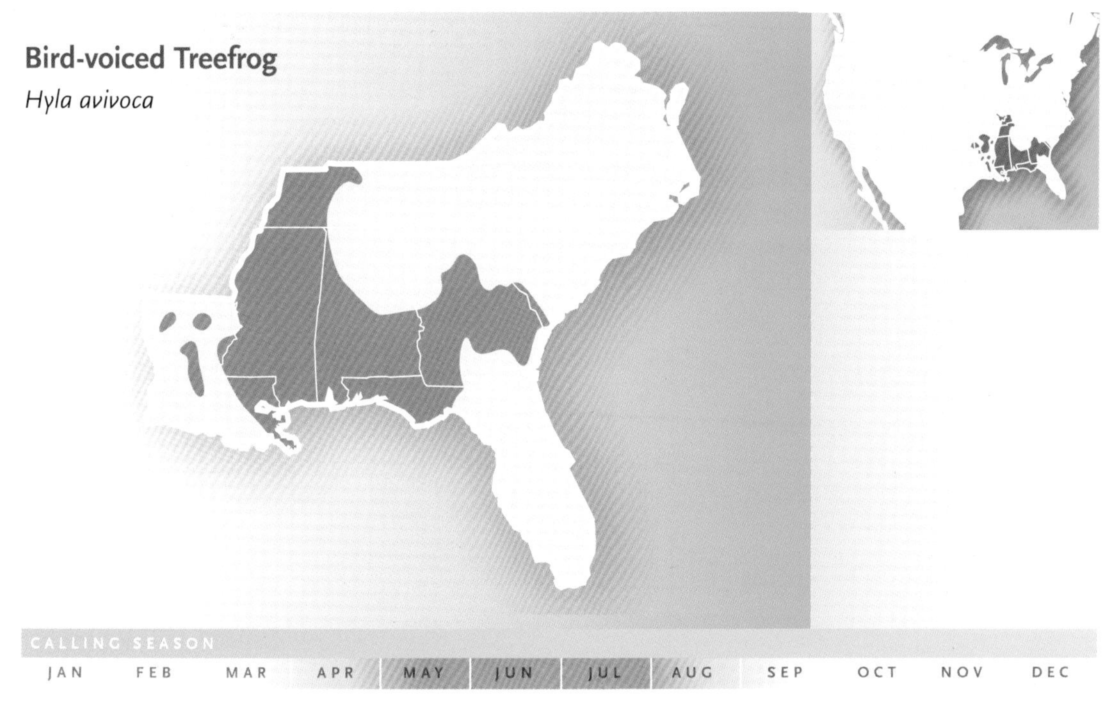

treefrog species, is a distinguishing character. Bird-voiced treefrogs might also be confused with squirrel treefrogs or pine woods treefrogs, but neither has a prominent rectangular blotch under the eye or the greenish coloration in the groin area.

Greenish spots on the thighs of bird-voiced treefrogs help to distinguish them from other species.

DISTRIBUTION AND HABITAT The geographic range extends patchily from southern Illinois southward to eastern Louisiana and eastward through all of Mississippi and the southern half of Alabama to the western part of the Florida panhandle, central Georgia, and southeastern South Carolina. Isolated populations exist in southeastern Oklahoma, central Arkansas, central and northern Louisiana, north-central Alabama, and north-central Georgia. Bird-voiced treefrogs are associated with large stream and river systems and their adjacent cypress–tupelo gum swamps. Breeding sites typically have large amounts of brush, such as buttonbush, that provides suitable calling sites for males. Outside the breeding season bird-voiced treefrogs usually retreat high into the tops of bald cypress, tupelo gum, or other trees but may occasionally be found under bark or hiding within dense vegetation at lower heights.

BEHAVIOR AND ACTIVITY These poorly studied, generally nocturnal frogs are active primarily during the late spring and summer. Males call at temperatures above about 70° F, often calling during the day from high in trees.

They probably spend their entire lives in or adjacent to cypress–tupelo gum river swamp habitats.

Bird-voiced treefrogs can be gray, brown, or green.

FOOD AND FEEDING Like most other treefrogs, bird-voiced treefrogs probably eat small to medium-sized insects such as moths and crickets. Recorded prey items include beetles and their larvae, moths, ants, and leafhoppers. Adults hunt primarily by ambush, perching on vegetation and waiting for potential prey to come within reach.

DESCRIPTION OF CALL The very distinctive, birdlike call consists of a series of rapidly repeated, high-pitched, whistle-like notes. The syncopation resembles the call of the pileated woodpecker, which often inhabits the same swampy habitats. A calling male approached by another calling male (or a human imitating a bird-voiced treefrog) will give one to three longer, harsh whistles, each lasting about 1 second, that presumably warn the other frog to find somewhere else to call.

REPRODUCTION Although males may call from high in the treetops before actual breeding begins, they descend and call from lower levels dur-

A bird-voiced treefrog calling

ing breeding periods. Calling usually begins at dusk and lasts until about midnight, or may terminate earlier if temperatures drop too low. Amplexus occurs in bushes above the water, and the female carries the male to the water, where she lays 400–800 eggs in small clusters. The eggs sink to the bottom or adhere to aquatic vegetation and hatch within 2 days. The tadpoles metamorphose after about a month.

Outside the breeding season bird-voiced treefrogs usually retreat high into the tops of trees.

PREDATORS AND DEFENSE Very little is known about predators of bird-voiced treefrogs, but a variety of birds, small to medium-sized mammals such as raccoons, and snakes such as watersnakes and juvenile rat snakes probably eat juveniles and adults. Tadpoles and eggs are probably eaten by a variety of aquatic predators such as insect larvae, turtles, and swamp-dwelling fish. The adults' primary defense is camouflage and remaining hidden high in trees whenever possible.

CONSERVATION Bird-voiced treefrog populations are susceptible to any disturbance that adversely affects their river and swamp habitats. Their actual status in many areas is unknown because their habitats are difficult to reach. The species is listed as Threatened in Illinois because only a few isolated populations have been documented, all from the southern part of the state.

The lack of any distinctive characteristics sometimes makes the squirrel treefrog difficult to identify.

How do you identify a squirrel treefrog?

SKIN
Smooth

LEGS
Relatively long and slender

FEET AND TOES
Moderate webbing on hind feet; large toe pads

BODY PATTERN AND COLOR
Brown to light green, sometimes with irregular spots

DISTINCTIVE CHARACTERS
None

CALL
Loud, nasal *waaak* repeated fairly rapidly

SIZE
max tadpole = 1.5"
typical adult = 1.25"

Squirrel Treefrog

Hyla squirella

DESCRIPTION Squirrel treefrogs are small and nondescript. The back is green to light brown, sometimes with irregular, indistinct spots. Most individuals have a faint, wavy light stripe running from the upper lip down along the body on each side. Some have a dark triangle or irregular spot between their eyes. The background color and presence of spots are subject to change depending on temperature and the animal's activity level.

WHAT DO THE TADPOLES LOOK LIKE? Squirrel treefrog tadpoles have a brownish green back and a yellowish underside. The sides are somewhat iridescent, and the middle of the belly is black. The tail, relatively long and very tall from top to bottom, has a mottled appearance.

SIMILAR SPECIES Because they lack distinctive characteristics, squirrel treefrogs are sometimes difficult to distinguish from other treefrogs. Perhaps the easiest way to identify a squirrel treefrog is

A squirrel treefrog tadpole

Squirrel Treefrog

Hyla squirella

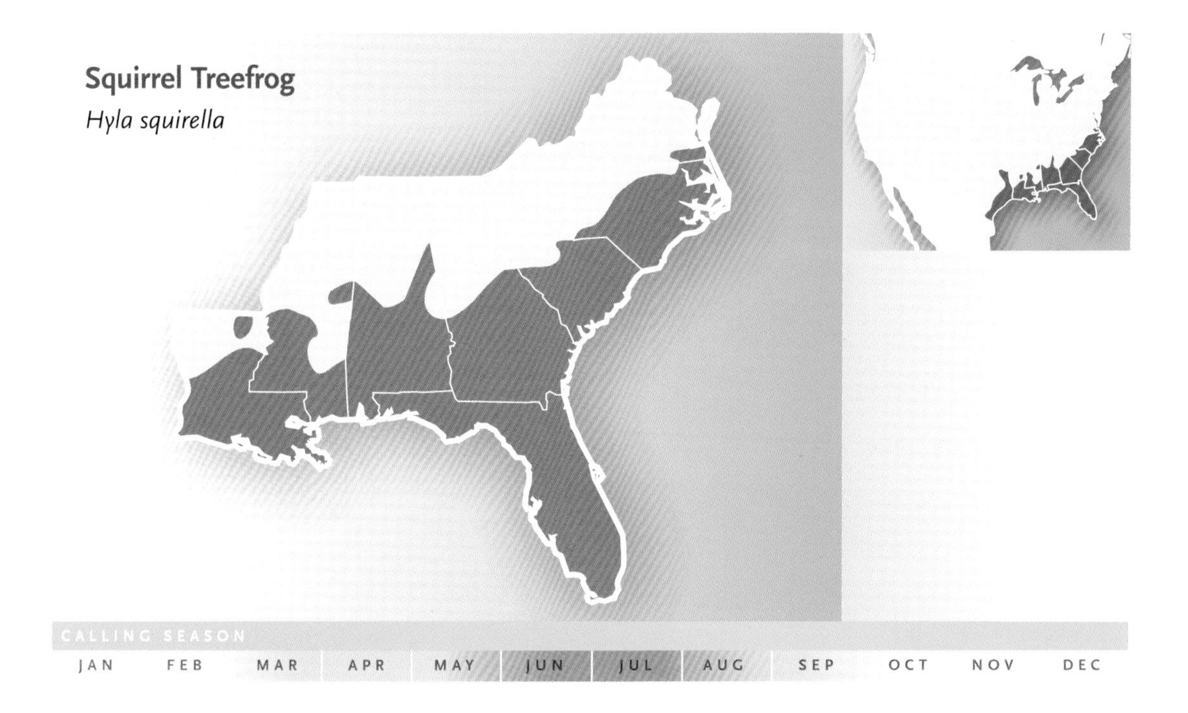

by the process of elimination. For example, the somewhat similar green treefrog has a well-defined ivory to yellowish stripe running along each side of the trunk. Green treefrogs are also larger and more streamlined than squirrel treefrogs and often have tiny orange spots on the back. Pine woods treefrogs have yellow spots on the dark posterior parts of their thighs.

Squirrel treefrogs are common throughout much of the southeastern Coastal Plain.

DISTRIBUTION AND HABITAT Squirrel treefrogs are found primarily in the Coastal Plain of the southeastern United States from southeastern Virginia south through all of Florida and westward to eastern Texas. They occupy a variety of habitats and will breed in almost any aquatic area, but they prefer fish-free wetlands, roadside ditches, and other shallow water bodies. They are often found on barrier islands, and are apparently fairly tolerant of salt water, having been reported breeding in pools with salinities approaching 50 percent. Squirrel treefrogs sometimes hitchhike along with humans and turn up in regions where they do not naturally occur, such as the Bahamas.

BEHAVIOR AND ACTIVITY Squirrel treefrogs are active primarily during the spring and summer, with the months of activity varying with latitude. Although generally nocturnal, they can often be heard calling from treetops or around houses during the day, especially after a drop in barometric pressure. Daytime retreats include loose bark and other sheltered places where the frog can rest on a vertical surface. They apparently spend the winter buried beneath soft soil or within rotting logs. Up to 40 individuals may overwinter communally under loose tree bark.

FOOD AND FEEDING Squirrel treefrogs hunt by ambushing insects and other small arthropods. They can often be observed foraging around lighted windows or on porches. Like many other treefrogs, they use their forelimbs to subdue unruly prey and force it into their mouth. Documented prey items include beetles, sowbugs (isopods), and ants.

DESCRIPTION OF CALL The advertisement call is a loud *waaak,* usually repeated several times, that resembles a duck quack. When rain is imminent, males often issue a somewhat different "rain call" that is reminiscent of a scolding squirrel; hence the common name.

REPRODUCTION Breeding occurs in late spring or summer, often coinciding with rain. Large choruses may form in freshwater habitats in coastal areas. Males frequently call while hidden within grass clumps in water only a few inches deep; they may also call from low perches around their pond or wetland, or even from the ground. Females lay 900–1,000 eggs singly or in small clusters, either on the bottom of the pond or attached to aquatic vegetation. Eggs hatch within a couple of days, and the tadpoles transform into the adult form within 1.5–2 months.

Squirrel treefrogs often call from high in the trees before afternoon thunderstorms.

PREDATORS AND DEFENSE Tadpoles are eaten by large aquatic insects such as diving beetles and dragonfly larvae. Ribbon snakes have been documented feeding on adults, and other species of semiaquatic snakes, birds such as night herons, and small to medium-sized mammals probably eat them as well. Some scientists have suggested that squirrel treefrogs may become prey of their larger relatives, Cuban treefrogs, in areas where their ranges overlap.

CONSERVATION Squirrel treefrogs are relatively abundant in many areas of the Southeast. Some scientists have reported declines in urban areas, but others have found these treefrogs to be abundant despite urbanization. Squirrel treefrogs are often the most common frog in minimally developed suburban areas. Road mortality may be a factor during migrations to breeding sites, but its effects on populations are unknown. Squirrel treefrogs are apparently introduced by human activities into regions where they do not naturally occur, but whether they become established or cause problems for native species is unknown.

Pine woods treefrogs are found throughout the southeastern Coastal Plain.

How do you identify a pine woods treefrog?

SKIN
Smooth

LEGS
Relatively long

FEET AND TOES
Moderate webbing on hind feet; large toe pads

BODY PATTERN AND COLOR
Brown or reddish brown, sometimes gray or grayish green; irregular markings on back

DISTINCTIVE CHARACTERS
White or yellow spots on posterior of dark thighs

CALL
Rat-a-tat-tat, somewhat like Morse code

SIZE
max tadpole = 1.5"
typical adult = 1.25"

Pine Woods Treefrog *Hyla femoralis*

DESCRIPTION These relatively small, slender treefrogs are usually brown or reddish brown with irregular large blotches on the back that resemble pine bark. Some individuals are gray or dull green. Background color varies from light to darker brown or gray depending on the individual's activity level, the temperature, or stress. Most adults have yellow to white, round or oval spots on the posterior portion of their thighs. The underside is white to cream.

WHAT DO THE TADPOLES LOOK LIKE? The tadpoles are dark green to brown above and have a light-colored belly and stripes on the tail musculature. The tail fins have black blotches on the edges and are somewhat reddish in between the blotches.

The tadpoles of pine woods treefrogs are brightly colored.

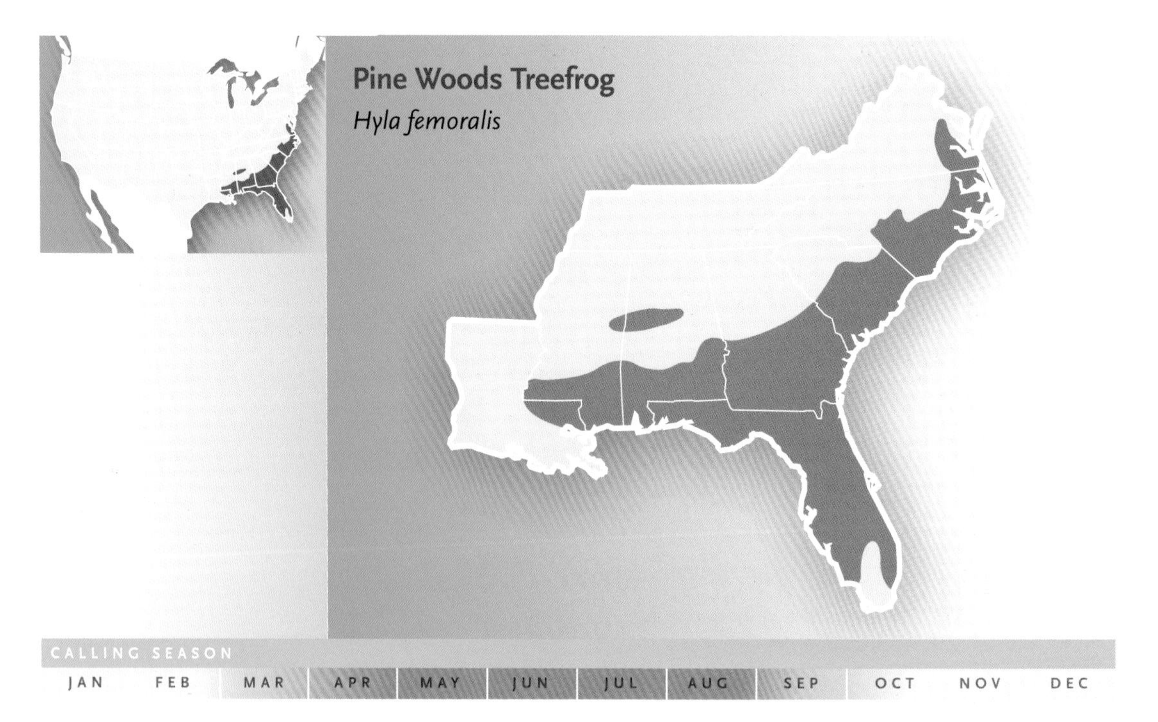

Most adult pine woods treefrogs have yellow to white spots on the posterior portion of their thighs.

SIMILAR SPECIES Pine woods treefrogs are probably most easily confused with squirrel treefrogs, but the latter lack well-defined light spots on the dark posterior of their thighs. Pine woods treefrogs might also be confused with gray treefrogs or bird-voiced treefrogs, but both of those species have fairly prominent white spots under their eyes and lack the circular light spots on the back of their thighs.

DISTRIBUTION AND HABITAT The geographic range encompasses most of the Coastal Plain of the southeastern United States, beginning midway up the Coastal Plain of Virginia and continuing southward through all of Florida except the Everglades and westward to eastern Louisiana. Pine woods treefrogs live primarily in forested areas, especially pine flatwoods, but may also occur in open habitats adjacent to forests. They typically breed in shallow forest pools and freshwater wetlands such as roadside ditches, Carolina bays, marshes, and shallow swamps. They generally do not breed in aquatic habitats containing predatory fish.

BEHAVIOR AND ACTIVITY These treefrogs can be found in large numbers at breeding sites or migrating to them, but are otherwise usually difficult to find, often staying hidden high in trees throughout the forest when not involved in reproductive activity. They may burrow in shallow, sandy soil and can sometimes be found hiding under loose bark. Pine woods treefrogs are rarely found during the day and are generally most active during the warmer months (April–September). They presumably remain dormant throughout most of the winter, taking refuge under logs or in shallow burrows, but have been known to emerge on warm days in southern latitudes.

FOOD AND FEEDING Pine woods treefrogs, like most treefrogs, feed predominantly on insects, which they capture by ambush. They are frequent visitors to back porches and windows where insects are attracted to lights. They are known to eat crickets and grasshoppers, beetles, ants, wasps, craneflies, moths, caddis flies, and jumping spiders.

The call of the pine woods treefrog has been said to resemble Morse code.

DESCRIPTION OF CALL The call resembles a telegraph operator sending random Morse code. Some observers think the call sounds more like a riveting machine or a *rat-a-tat-tat*. Regardless of the comparison, the call is quite distinctive and once heard is unlikely to be confused with the call of any other species. During summer days, pine woods treefrogs often give a similar but briefer call from high in the tops of trees.

REPRODUCTION Males begin calling in late March or early April and may continue through the summer, even into early fall in some locations. They typically call from vegetation in and adjacent to breeding sites. Large choruses are common after heavy rains, but some individuals may call even during dry periods if the wetland or pond still holds water. Eggs are laid as sticky clumps attached to vegetation only an inch or so below the water's surface. Females may lay up to 2,000 eggs, usually in separate clusters of 100 or so each. Eggs hatch in late spring or early summer, and the tadpoles usually metamorphose in 1.5–2.5 months.

PREDATORS AND DEFENSE Known predators of adults include numerous species of snakes, such as southern banded watersnakes, racers, juvenile rat snakes, garter snakes, and ribbon snakes. Mammals such as raccoons and wading birds are also likely predators. Various turtles, snakes, and aquatic insects eat the tadpoles.

CONSERVATION Pine woods treefrogs can be locally common and are not considered threatened in any part of their range. Destruction or contamination of wetland habitats would likely reduce populations, and in some places (e.g., areas of high urbanization) this species is probably locally extinct. Extensive cutting of pine forests would likely also be detrimental.

Cuban treefrogs can get much larger than treefrogs native to the Southeast.

How do you identify a Cuban treefrog?

SKIN
Mostly smooth with a few scattered warts

LEGS
Long

FEET AND TOES
Toes of hind feet partially webbed

BODY PATTERN AND COLOR
Highly variable green, gray, or brown; white belly

DISTINCTIVE CHARACTERS
Large size, large toe pads; skin fused to top of head

CALL
Raspy squawk occasionally followed by clicks

SIZE
max tadpole = 1.25"
typical adult male = 3"
typical adult female = 4.5"

Cuban Treefrog

Osteopilus septentrionalis

DESCRIPTION The body is usually solid green, gray, or brown; darker blotches or short stripes may be present on the back, sides, and legs. The skin on the body has widely scattered small warts. The belly is white and granular. The rounded, flat toe pads are disproportionately larger than those of any other southeastern frog, including the other treefrogs. Partial webbing is apparent between the toes of the hind feet. Some young have a white or yellowish side stripe, but it does not run through or beneath the eye. The large size is a notable feature; the largest males reach lengths of about 3 inches, and some females are more than 5 inches long.

WHAT DO THE TADPOLES LOOK LIKE? The tadpoles typically have a round body that is brownish or black above and lighter below. The pointed tail fin is transparent and has a grayish brown tint and scattered dark spots. The tadpoles often swim in schools and are usually 1–1.25 inches in total length.

SIMILAR SPECIES Our native treefrogs are the species most likely to be confused with the Cuban treefrog in the Southeast. The large size, extra-large toe pads, lack of a stripe on the sides that continues forward through or under the eye, and slightly webbed hind feet separate this species from the others. The Cuban treefrog is also the only southeastern treefrog whose skull is fused with the skin above it. This unique trait, which makes the top of the head rough and scratchy, is apparent in all but the smallest individuals.

DISTRIBUTION AND HABITAT These treefrogs, native to Cuba and other West Indies islands, were reported in the 1930s from the Florida Keys, where they apparently had been resident for many years. They were presumed to have arrived there in commercial shipping, but some amphibian biologists believe their arrival could have occurred naturally. These treefrogs were present on the Florida peninsula by the 1950s, and by the early 2000s were occurring in increasing numbers in northern Florida and the panhandle, which they almost certainly reached through human transport. They are presently abundant throughout much of the Everglades. No established populations had been reported outside Florida by the early 2000s, but two authenticated reports of individual Cuban treefrogs have come from Savannah, Georgia. Cuban treefrogs can be found in various woodland habitats but are most abundant and most apparent around human habitations, especially large heated build-

The Cuban treefrog's toe pads are disproportionately larger than those of any other southeastern frog.

Cuban Treefrog

Osteopilus septentrionalis

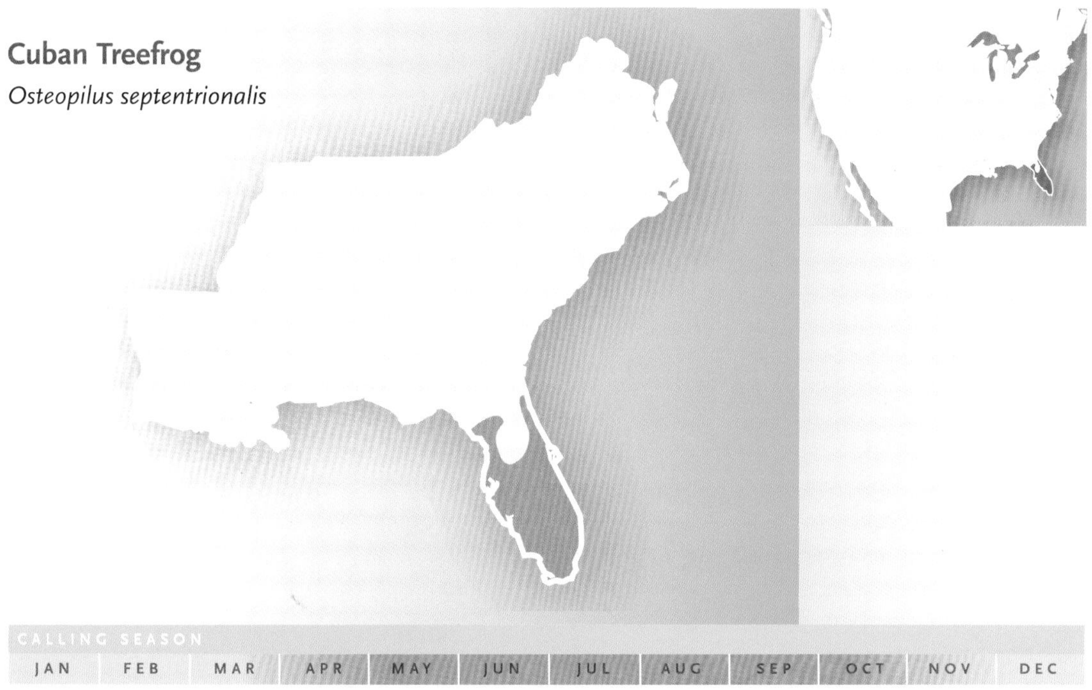

Cuban treefrogs vary in color from gray to brown to green.

Cuban treefrogs in amplexus

ings that have warm, moist hiding spots; outdoor lighting that attracts insects and other prey; and standing water in the vicinity for breeding.

BEHAVIOR AND ACTIVITY Cuban treefrogs require warm temperatures for activity. They become torpid and completely inactive at temperatures below about 50° F. They are active from dusk to dawn, and sometimes during the day on dark, overcast days, although they generally retreat to hiding areas at daybreak. Large migrations of adults from feeding areas to breeding sites occur following heavy rains during the breeding season.

FOOD AND FEEDING Cuban treefrogs are ambush predators that eat invertebrates, including beetles and large roaches, and a wide variety of vertebrates including lizards such as brown anoles, Indo-Pacific geckoes, and tropical house geckoes; small snakes such as the Florida brown snake; frogs such as green and squirrel treefrogs, southern leopard frogs, and smaller individuals of their own species; and toads such as southern toads and eastern narrowmouth toads. The tadpoles primarily eat algae but have also been known to eat other tadpoles, including members of their own species.

DESCRIPTION OF CALL The advertisement call sounds like a rasping version of a cross between an eastern spadefoot toad's incessant *qwaah* and a southern leopard frog's "rubbed balloon" croak.

REPRODUCTION Cuban treefrogs breed in every month in southern Florida, but breeding peaks during the wet season, which can last from May through October. After heavy rains, males form large choruses that attract females to aquatic habitats—most commonly fish-free ditches and retention ponds, swimming pools, and flooded pools alongside lakes. They have also been reported to breed in brackish habitats. Each female produces more than 3,000 eggs, laying them in batches of 75–1,000 that float on the surface like a mat. Temperatures above 80° F are crucial for successful breeding and tadpole development. The eggs hatch in a little more than a day, and if the water is as warm as 95° F, the tadpoles undergo metamorphosis in as few as 3 weeks. If the water is below 85° F, metamorphosis can take more than 3 months. Cuban treefrogs transform into froglets when they are 0.5–0.75 inch long.

PREDATORS AND DEFENSE Known predators of adults in Florida include native snakes (rat snakes, racers, garter snakes, ribbon snakes) and birds (barred owl, crow) as well as the knight anole, an introduced lizard. Crows capture and eat tadpoles as well as recently metamorphosed juveniles at the margins of breeding sites. The skin of adults secretes defensive compounds that are highly toxic to humans and probably some potential mammalian predators as well. The secretion is particularly painful if it comes in contact with the eyes, mouth, or nasal membranes. Once captured by a predator, these frogs inflate their body and sometimes give a loud scream.

CONSERVATION Conservation initiatives that involve this species are intended not to protect the Cuban treefrog, but rather to protect native species, especially other treefrogs, from the negative impacts this exotic species can have on them. As both a predator of other treefrogs and a potential competitor for food, the Cuban treefrog is a confirmed threat in areas of Florida where it has become established. Conservation concern is heightened by the ease with which these frogs are inadvertently dispersed by people to potential new sites, sometimes repeatedly, and their ability to succeed in a wide range of habitats.

Did you know?

Some frogs produce a loud, drawn-out scream when attacked by a predator. The scream is given with the mouth open and may attract other animals that may distract the predator.

Partial webbing is visible between the toes of this Cuban treefrog's hind feet.

TRUE FROGS

A leopard frog from South Carolina

How do you identify a southern leopard frog?

SKIN
Smooth

LEGS
Muscular and long

FEET AND TOES
Hind feet extensively webbed

BODY PATTERN AND COLOR
Body green or brown; rounded dark spots on back; belly white

DISTINCTIVE CHARACTERS
Prominent dorsolateral ridges; rounded dark spots on the back; whitish spot on center of tympanum; white belly

CALL
Varies from rapid "chuckling" to guttural croaking

SIZE
max tadpole = 3.25"
typical adult = 3"

Southern Leopard Frog *Rana sphenocephala*

DESCRIPTION Southern leopard frogs are generally slender with smooth skin, a pointed head, and webbing on the hind feet. Light dorsolateral ridges extend from behind the eye along either side of the back, and a thick, light line runs along the upper jaw. The body may be light brown or green above, and both colors are present on some individuals. Irregular round or oval dark spots are visible on the back and sides and are irregularly scattered on most individuals. The belly is white with no markings. The tympanum typically has a light spot in the center. Southern leopard frogs commonly reach a length of 3.5 inches, and some individuals approach 5 inches; females grow considerably larger than males.

WHAT DO THE TADPOLES LOOK LIKE? Southern leopard frog tadpoles vary in body color from dark green to brownish and may have dark markings. The tail is translucent in some individuals or may have brownish spots or mottling. The belly is light, usually pinkish, and a white line is often apparent between the nostril openings on the snout. The tadpoles reach a length of about 2.5 inches.

A leopard frog tadpole from Florida

Leopard frogs are variable in color, ranging from brown to green.

Some leopard frogs have few or no spots.

SIMILAR SPECIES The prominent dorsolateral ridges and rounded dark spots on the back readily distinguish the southern leopard frog from most other frogs; pickerel frogs have similar markings, but their spots are squarish and usually occur in pairs. Also, pickerel frogs have yellow in the groin area and on the underside of the thighs.

DISTRIBUTION AND HABITAT Southern leopard frogs are found throughout every southeastern state except for the mountainous areas of Tennessee, North Carolina, and Virginia. The overall geographic range extends west to central Texas, north to southeastern Iowa and central Illinois, and up the East Coast to southeastern New York. These frogs are usually abundant wherever water is present. Lakes, reservoirs, farm ponds, marshes, river swamps and floodplains, creek margins, ditches, Carolina bays, and any other aquatic site—as well as peripheral areas with woods or other shrubby vegetation—can be home to southern leopard frogs. Individuals have even been observed in slightly brackish waters in coastal areas. During the warm months, adults may become terrestrial in moist areas with heavy ground vegetation.

BEHAVIOR AND ACTIVITY Southern leopard frogs are active in all months in most of the Southeast but become dormant during extremely cold periods. They move around mostly at night but will bask along the margins of ponds, often where tall grass or other vegetation provides camouflage, on sunny days. The species is relatively terrestrial, moving between breeding

Southern Leopard Frog

Rana sphenocephala

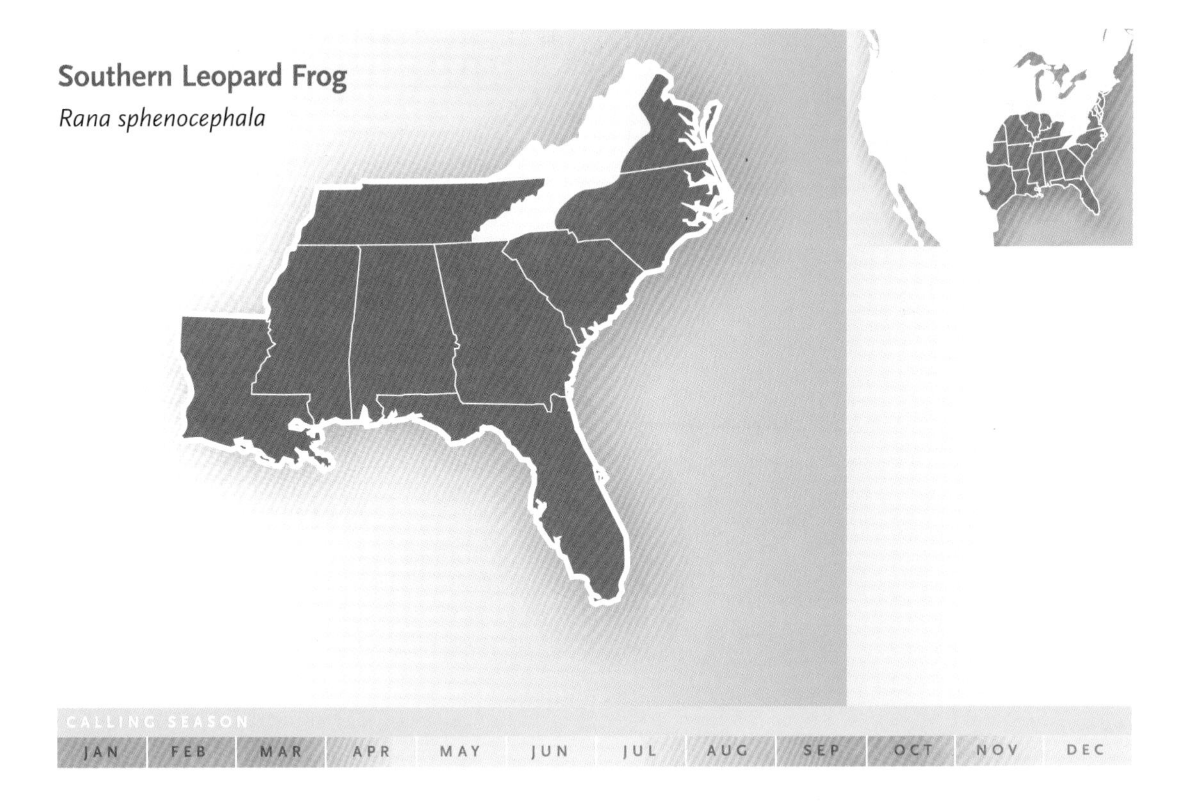

sites and habitats used for feeding or for refuge from cold or drought. During years of successful breeding in some southeastern wetlands, recently metamorphosed juveniles may emerge into the surrounding terrestrial areas by the tens of thousands over a several-day period.

A male leopard frog waiting for a female

FOOD AND FEEDING Southern leopard frogs eat a variety of insects, both terrestrial and aquatic, and are also known to eat crayfish. Tadpoles eat diatoms, green algae, and small aquatic plants.

DESCRIPTION OF CALL The advertisement call is highly variable but is usually some combination of rapid chuckles and croaks that sound like a rubber balloon being twisted. From a distance, large choruses create a low roar and calls of individuals may be difficult to discern. At colder temperatures, the chuckling portion of the call can resemble the call of a wood frog. A male grasped by another frog, or even by a human, gives a release call that has been described as a repeated guttural chuckle. Leopard frogs captured by a snake or other predator sometimes utter a high-pitched alarm call that is best described as a scream. Unlike their other calls, this call is given with an open mouth.

REPRODUCTION Southern leopard frogs leave their terrestrial habitats for aquatic breeding sites in response to rain. As long as temperatures are moderate, they will breed in all months of the year in the Southeast, but more from fall to spring than in midsummer. Females lay more than 1,000–1,500 eggs in still, shallow water. The nearly spherical egg masses float near the surface and are usually attached to vegetation. Several females may lay their eggs together in clusters in the same part of a wetland. In the Southeast, tadpoles take about 3 months to metamorphose, typically at a length of about 0.75–1.25 inches. Because the adults breed throughout the year, young frogs may emerge from wetlands during any month, but most do so in early summer.

The eggs of leopard frogs are laid in a round mass about the size of a softball.

PREDATORS AND DEFENSE Known predators include raccoons; birds, including great blue herons and grackles; and several species of snakes, including watersnakes, cottonmouths, and racers. Southern leopard frogs are particularly adept at escaping by jumping rapidly in unpredictable directions on land or swimming into the debris and vegetation at the bottom of a pond.

CONSERVATION Southern leopard frogs are common in most parts of their geographic range. They are not considered in need of protection in any southeastern state, although some populations may have been overcollected for fish bait and for biological supply houses that sell them for classroom specimens and research.

COMMENTS The scientific name of this species and the nature of its relationship to close relatives have been unsettled since the 1800s. By the late 1990s, most herpetologists accepted the species name *Rana sphenocephala*, and some have accepted two subspecies that are impossible to distinguish based on external characters. The Florida leopard frog (*Rana sphenocephala sphenocephala*) is primarily restricted to peninsular Florida. The other subspecies, the southern leopard frog, is *R. s. utricularia*. In 2006, amphibian biologists placed this species in the genus *Lithobates* and now refer to it as *Lithobates sphenocephalus*, with subspecies *L. s. sphenocephalus* and *L. s. utricularia*.

A pickerel frog from the mountains of North Carolina

How do you identify a pickerel frog?

SKIN
Smooth

LEGS
Long and muscular

FEET AND TOES
Hind feet extensively webbed

BODY PATTERN AND COLOR
Brown or tan with squarish dark spots in two rows

DISTINCTIVE CHARACTERS
Dark spots are squarish; thick dorsolateral ridges; yellow-orange on groin and rear of thigh

CALL
Relatively subdued snore

SIZE
max tadpole = 3"
typical adult = 2.5"

Pickerel Frog

Rana palustris

DESCRIPTION Pickerel frogs are smooth-skinned, medium-sized frogs with robust hind legs and extensively webbed toes. The back is greenish brown to tan. A half-dozen or more large, square spots form two rows down the center of the back between the thick dorsolateral ridges, with a smaller number of single squares on the sides beneath the folds. The spots sometimes fuse to form long rectangles or thick lines. The belly varies from white to mottled gray to dark gray. The groin and rear part of the thigh, which is hidden when the frog is sitting, is yellow or orange. A prominent white line is present on the upper jaw. A vocal pouch is present on each side of the throat. Breeding males have greatly enlarged thumbs. Adult females are larger than adult males, often by more than an inch.

WHAT DO THE TADPOLES LOOK LIKE? The tadpoles reach a total length of about 3 inches and are brownish or greenish on the back with a mix of fine yellow and black dots. The belly is pale yellow or cream colored. The large tail fin is dark but may be lightly speckled.

This pickerel frog tadpole is already starting to display its spots.

Pickerel Frog

Rana palustris

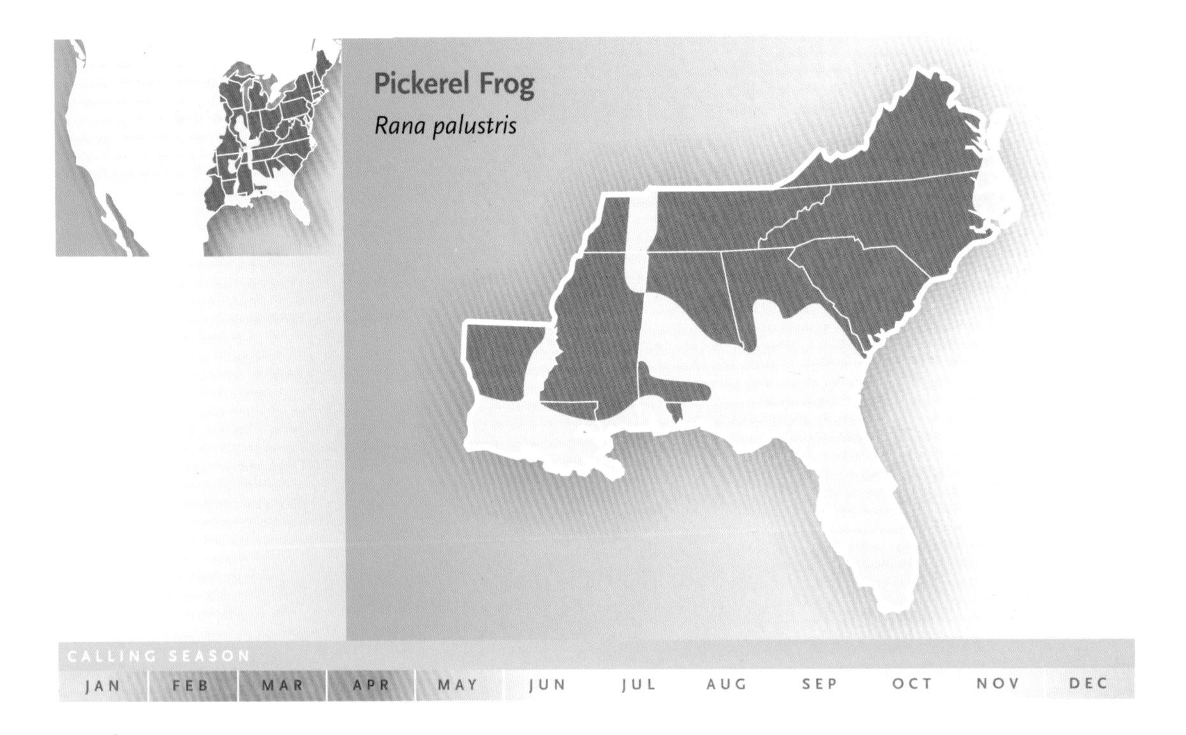

Their yellow thighs and groins distinguish pickerel frogs from leopard frogs.

SIMILAR SPECIES Pickerel frogs are more likely to be confused with leopard frogs than any other native species. The dark spots on the back of pickerel frogs are square instead of rounded, however, and are usually in two rows down the back between the dorsolateral ridges. In addition, no other true frog in the Southeast has yellow or orange on the groin and rear thigh.

DISTRIBUTION AND HABITAT Pickerel frogs range from southeastern Canada through all of New England, west through Wisconsin, and south to eastern Texas. They are present in all of the southeastern states, although in only one county (Escambia) in Florida. In many parts of the Southeast, the geographic distribution is not continuous, and the species is absent—or at least unreported—from many localities, especially in the Coastal Plain of Alabama and Georgia. Pickerel frogs are associated with diverse aquatic habitats that appear to vary regionally. For example, the species is common within floodplain swamps in Congaree National Park in South Carolina, but apparently does not occur in similar habitats in Alabama and Georgia. Pickerel frogs have been associated with trout streams; clear waters in flat, upland forests; bogs and grassy, open meadows; farm ponds; and aquatic habitats at the openings of caves.

BEHAVIOR AND ACTIVITY Pickerel frogs may be active during the day but become nocturnal during the summer. They spend winter buried in mud

Pickerel frogs are found throughout much of the eastern United States.

or debris at the bottom of ponds, swamp pools, or slow-moving water along stream edges. Some aspects of the general ecology and life history, including size of the home range, whether they establish territories, and life span, are completely unknown.

The usually square spots arranged in two rows help distinguish pickerel frogs from leopard frogs.

FOOD AND FEEDING Pickerel frogs eat small invertebrates, primarily insects and spiders as well as mollusks such as snails. As is true for most members of this family, nearby movement of any small animal usually elicits a rapid feeding response.

DESCRIPTION OF CALL The advertisement call has been described as a relatively low-pitched, croaky snoring or grinding sound. It is not exceptionally loud and cannot be heard at great distances.

REPRODUCTION Males in the southernmost populations begin calling as early as December; those in more northern areas of the Southeast begin in March or April. Mating usually ends by May. Males generally call from shallow water and sometimes even from beneath the surface. Adults of both sexes immigrate to ponds, flooded marshlands, seepage pools around streams, and isolated wetlands for breeding, and afterward disperse into surrounding woodland habitats or wetland margins. Each female lays her eggs—as many as 2,500–3,000—in spherical clusters that she attaches to surface or submerged vegetation. The eggs hatch into tadpoles after

1.5–3.5 weeks, and the tadpoles begin metamorphosis within 2–3 months after hatching. Newly metamorphosed froglets are 0.75 to just over 1 inch long.

PREDATORS AND DEFENSE Bald eagles and minks are known to prey on adults, and a variety of other vertebrate predators, including bullfrogs, snakes, raptors, and raccoons, probably eat them as well. Known or potential predators of tadpoles include fish, red-spotted newts, salamander larvae, dragonfly naiads, and diving beetles. Reports that pickerel frogs secrete toxins that make them distasteful to predators have generated debate. Some herpetologists say that the skin of pickerel frogs is so toxic that most snakes will not eat them and other frog species placed in the same collecting bag will die, but others view such claims as overstated. It is possible that both groups are correct; that is, pickerel frogs may produce toxic skin secretions in some parts of their range but not others, or may be toxic only at certain life stages. A third possibility is that pickerel frogs produce toxic secretions only in some habitats or under certain diet regimes.

CONSERVATION Limited research on pickerel frogs in the Southeast coupled with the species' spotty distribution in a variety of habitats have led to uncertainty about which environmental alterations are most likely to affect distribution and abundance in any given region. Their apparent rarity in some areas could make some populations particularly sensitive to certain environmental disturbances or contamination. The species is noted for crossing highways in large numbers when migrating to wetlands, and building roads across breeding routes is likely to result in high mortality in some populations.

Pickerel frogs' hind feet are extensively webbed.

COMMENTS The subspecies *R. p. mansuetti* was described from the Atlantic Coastal Plain, but later studies revealed that the observed variation in pattern was not restricted to a geographic region, and the subspecies is no longer recognized. In 2006, amphibian biologists placed this species in the genus *Lithobates*.

Green frogs can be found in a variety of water bodies.

Green Frog, Bronze Frog

Rana clamitans

DESCRIPTION Green frogs are medium-sized true frogs (usually no more than 3 inches long in the Southeast) that come in various shades of greens and browns; many individuals are part green and part brown. Individuals with a brown body often have a green head and/or jaws. One of the two recognized subspecies, the bronze frog (*R. clamitans clamitans*), is typically light brown or bronze. Some individuals have no markings on the back and sides while others have irregular dark markings or small spots. The belly is mostly mottled white, although the underside of the chin and throat may be smoky gray. The underside of the chin of some adult males is yellowish. The dorsolateral ridges are prominent on the front part of the back but usually become indistinct toward the rear. The skin is mostly smooth but is covered with small, wartlike protuberances. The hind feet have extensive webbing between the toes.

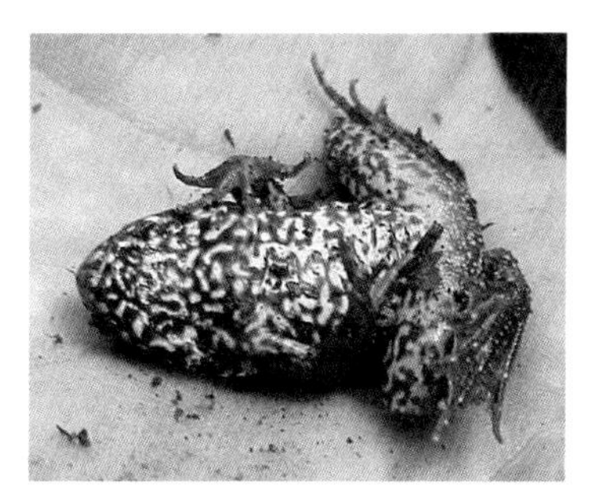

The belly of the green frog is usually mottled.

WHAT DO THE TADPOLES LOOK LIKE? Like those of most other true frogs, green frog tadpoles are relatively robust and vary in body color from green-

How do you identify a green frog?

SKIN
Mostly smooth

LEGS
Muscular and long

FEET AND TOES
Hind feet webbed

BODY PATTERN AND COLOR
Back usually unmarked, greenish to brown; belly mostly white

DISTINCTIVE CHARACTERS
Dorsolateral ridges extend only partway down the body

CALL
Single *glunk* sounding like a loose banjo string being plucked

SIZE
max tadpole = 4"
typical adult = 3"

A bronze frog

Green frogs often have spots on their backs.

ish to brown. Some individuals have numerous small, dark spots. The tail fin may be translucent or spotted. The belly is whitish. Tadpoles generally range between 1.25 and 3.5 inches in length.

SIMILAR SPECIES Green frogs are most easily confused with bullfrogs and pig frogs but can be distinguished by their prominent dorsolateral ridges, which extend past the tympanum. They can be separated from other true frogs because these ridges usually end about two-thirds of the way down the back rather than running the entire length of the body. The lack of conspicuous spotting also differentiates green frogs from several of the true frog species, such as leopard and pickerel frogs.

DISTRIBUTION AND HABITAT Green frogs are found throughout every southeastern state except Florida, where they are present only in the northern half. Of the two subspecies, the bronze frog (*R. c. clamitans*) has a more southerly distribution; the green frog (*R. c. melanota*) is found throughout Virginia, most of North Carolina, and eastern Tennessee to the northeastern corner of Mississippi and the northern portions of Alabama, Georgia, and South Carolina. Green frogs occur in a variety of aquatic habitats in the Southeast but are generally associated with permanent bodies of water, including farm ponds, lakes and reservoirs, swamps and freshwater marshes, and the margins of rivers and streams. They are frequently found on land, usually in damp areas, but typically near water. They sometimes venture farther away from water during or after rains.

BEHAVIOR AND ACTIVITY Adults and juveniles are active both day and night and are usually found in the vicinity of permanent water into which they can jump when threatened. Adults have an activity area that averages less than an acre, but juveniles have been known to travel 3 miles away from the pond where they were born. Green frogs commonly hibernate underwater at sites that do not freeze, including flowing water, but sometimes spend cold periods beneath leaves or other ground litter. Unlike many other species of frogs, green frogs actively try to escape when captured.

FOOD AND FEEDING Green frogs eat beetles, flies, butterflies, dragonflies, and other insects and an array of other invertebrates including worms, crayfish, spiders, millipedes, and centipedes. Smaller frogs, snails, and slugs are sometimes on the menu as well. The tadpoles typically eat algae and plant detritus but sometimes also eat the eggs of other frogs.

DESCRIPTION OF CALL The loud advertisement call can sometimes be heard during the day as well as at night and is unlikely to be confused with other frog calls. It resembles the nonmusical twang of a single banjo note or a tightened rubber band, and each succeeding note is distinct from the previous one. Individual males often call from solitary locations that

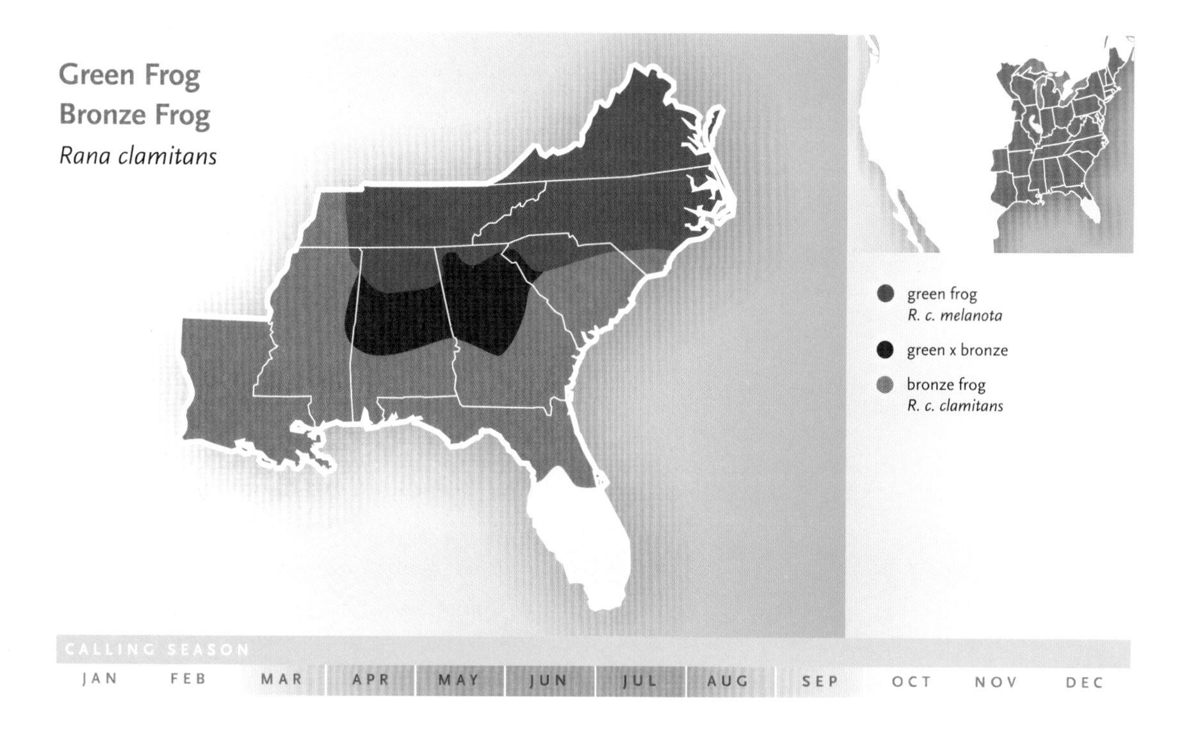

are distant from other calling males. The call is easily imitated, and a person with patience can elicit an aggressive response call from a male. The response call—a loud and deliberate *glunk* followed by a low, drawn-out, syncopated guttural sound that is similar to the call of a bullfrog (although higher pitched)—is assumed to be a warning to other males entering a calling male's territory. When disturbed on land, individuals often leap into the water while simultaneously giving an abrupt *eeek*.

REPRODUCTION Green frogs in the Southeast mate in permanent aquatic habitats as early as March and as late as September. Males establish territories and defend them from other male green frogs with warning calls, aggressive posturing, and even physical combat. The size of a male's territory varies according to the amount of aquatic and shoreline vegetation, with males being closer (within 3 feet) to one another when plants are thick but spaced more than 10–15 feet apart when vegetation is sparse or absent. A single female may lay as few as 1,000 or as many as 7,000 eggs in shallow vegetated areas along the shoreline. The eggs float on the water's surface and can form a thin sheet up to a foot wide. Green frogs can develop from the egg through the tadpole stage in 3 months, but can slow their development during autumn in order to spend the winter in the tadpole stage. The tadpoles typically reach lengths between 2 and 3.5 inches; recently metamorphosed froglets are less than 1.5 inches long.

The male green frog can be distinguished from the female by its large tympanum.

PREDATORS AND DEFENSE Green frogs are eaten by a variety of predators that live in or around wetlands, including bullfrogs, watersnakes, garter snakes, ribbon snakes, ducks, and herons. Hawks and crows take frogs that venture away from the water, and adult green frogs will cannibalize smaller members of their own species. Tadpole predators include aquatic insects such as dragonfly naiads, diving beetle larvae, and giant water bugs. Many turtles will eat the eggs of green frogs. The primary means of defense for juveniles and adults is to jump into the water and hide among the debris on the bottom.

CONSERVATION Because they have a wide geographic range and are common around most permanent bodies of water, green frogs are not in any danger on a broad scale, although local populations can be affected by habitat disruption resulting from peat mining or removal of shoreline vegetation.

COMMENTS In 2006, amphibian biologists placed this species in the genus *Lithobates*.

A wood frog from Blacksburg, Virginia

How do you identify a wood frog?

SKIN
Mostly smooth

LEGS
Long and muscular

FEET AND TOES
Hind feet moderately webbed

BODY PATTERN AND COLOR
Brown to red; belly white

DISTINCTIVE CHARACTERS
Broad, dark, masklike streak from nose to behind eye; distinct dorsolateral ridges

CALL
Continual raspy quack

SIZE
max tadpole = 2.25"
typical adult = 2.25"

Wood Frog

Rana sylvatica

DESCRIPTION Most wood frogs are a uniform reddish brown or tan, but some individuals may be dark brown to almost black. The back often lacks spots. A large dark area in front of and behind the eye resembles a mask. A distinct white or yellowish line runs the length of the upper jaw. Dark crossbars are apparent on the upper surface of the legs of light-colored individuals. The belly is often completely white and unmarked except for a dark spot on either side of the chest. The skin is generally smooth but may be slightly rough in some individuals. The dorsolateral ridges are very prominent and usually lighter in color than the back. The hind feet are noticeably webbed. Adult males are smaller than adult females.

WHAT DO THE TADPOLES LOOK LIKE? Wood frog tadpoles are brown or greenish brown and sometimes have bronze or gold flecks. Older tadpoles have a dark upper lip bordered by a light line. The tail is translucent. Tadpoles metamorphose at lengths of 2 to slightly over 2.25 inches.

The dark face mask distinguishes the wood frog from most other species.

SIMILAR SPECIES Wood frogs' prominent dorsolateral ridges, lack of spotting on most of the body, and dark mask that passes through the eye from the snout to the back of the jaw will distinguish them from other true frogs within their geographic range.

Some wood frogs have heavy black spotting.

Wood frogs range from brown to reddish brown in color.

DISTRIBUTION AND HABITAT Primarily a northern species, the wood frog is the only American frog found north of the Arctic Circle. In the Southeast, wood frogs are found primarily at higher elevations in the colder mountainous regions of the Appalachians in Alabama, Georgia, the Carolinas, and Tennessee. They also occur in some parts of the Piedmont and upper Coastal Plain in Virginia, and in two counties (Hyde and Tyrrell) in the Coastal Plain of North Carolina. As the common name implies, except for the relatively short breeding season, wood frogs are typically associated with wooded areas, primarily deciduous forests with a ground cover of leaves. Individuals often occupy terrestrial habitats hundreds of yards away from the small, isolated ephemeral wetlands where they breed.

BEHAVIOR AND ACTIVITY Adults are most often encountered en route to or from breeding sites or during the mating period at a pond itself. Overland movements of a half-mile up to almost a mile have been reported. After spending at most a few days at breeding ponds, the adults depart to spend the remainder of the year on land, usually in moist forests. Wood frogs hibernate in shallow burrows and can tolerate extremely cold weather because they produce "antifreeze" compounds that permit them to withstand subfreezing conditions without damage to their cells. Individuals can tolerate freezing of up to 70 percent of their body water.

FOOD AND FEEDING Wood frogs probably eat mostly insects, including beetles and flies, as well as spiders and other small invertebrates. The tad-

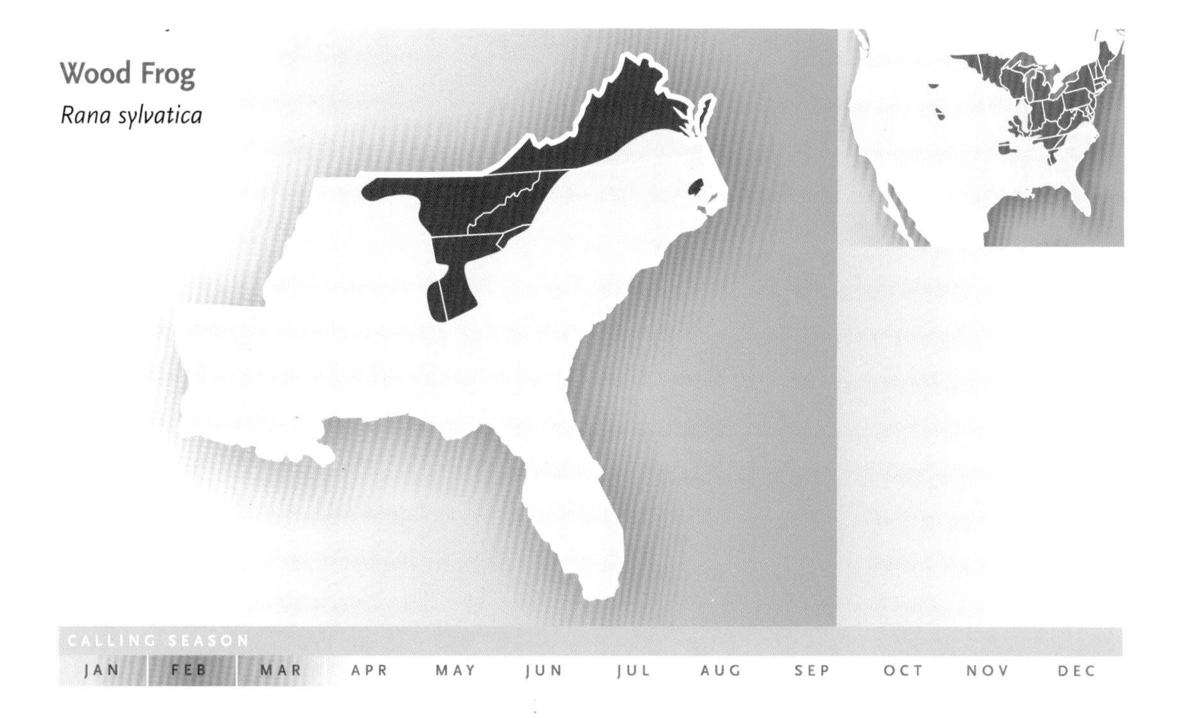

poles are particularly noteworthy for consuming not only algae and other plant materials in the water but also for eating the eggs and larvae of toads and salamanders. They are sometimes cannibalistic as well.

DESCRIPTION OF CALL The advertisement call has been variously described as *wurrk, wurrk, wurrk, craw-aw-auk,* and like a duck quacking. The call is lower in volume than most other frog calls and cannot be heard from as far away.

REPRODUCTION Wood frogs epitomize "explosive breeders." Virtually all of the adults in a population show up at the breeding site within a week's time. Wood frogs are among the earliest breeders throughout much of their geographic range. In the Southeast, most mating occurs in January or February, often when nights are cold and the breeding pond margins are still frozen. Individuals typically return to the same breeding site in subsequent years. Each female lays 300–1,500 eggs in clusters just below the water's surface in the same area of the breeding pond as other females, a behavior known as communal nesting. The tadpole stage lasts 2–4 months depending in part on springtime water temperatures.

PREDATORS AND DEFENSE Adults are prey for a variety of forest predators, including mammals such as raccoons, skunks, and foxes; large birds such as owls; and terrestrial snakes such as racers, watersnakes, and garter snakes. The larvae of pond-breeding salamanders and leeches eat the

Did you know?

Wood frogs are found above the Arctic Circle and can survive freezing of most of the water in their bodies.

Egg masses of wood frogs

Wood frogs begin breeding in shallow wetlands in late winter or early spring. This pair is getting ready to lay eggs.

eggs and tadpoles. Fish can also be voracious predators on the eggs and tadpoles, but wood frogs generally lay their eggs in fish-free aquatic areas.

CONSERVATION Wood frogs are generally common throughout much of their wide geographic range. They are less common in many southeastern habitats, probably because they are at the edge of their natural range there and conditions suitable for breeding may not occur every year. Due to the long distances adults sometimes travel between their aquatic breeding sites and terrestrial areas where they spend most of the year, some populations are especially susceptible to highway mortality. The protection of small temporary wetlands could be critical to the persistence of wood frog populations in some areas.

COMMENTS In 2006, amphibian biologists placed this species in the genus *Lithobates* and now refer to it as *Lithobates sylvaticus*.

A carpenter frog from Pender County, North Carolina

How do you identify a carpenter frog?

SKIN
Smooth

LEGS
Moderately stout and long

FEET AND TOES
Webbing on hind feet not complete

BODY PATTERN AND COLOR
Brown above with four light lines extending down the back and sides

DISTINCTIVE CHARACTERS
Narrow head; no dorsolateral ridges; four stripes extending along the body

CALL
Like a pair of hammers hitting a board rapidly; repeated two to four times

SIZE
max tadpole = 3.5"
typical adult = 2"

Carpenter Frog *Rana virgatipes*

DESCRIPTION Carpenter frogs are smaller than most other true frogs; adults do not exceed 3 inches in length. The back is dull brown with dark flecking in some individuals, and four yellowish to light brown stripes extend the length of the back and sides. The light-colored belly usually has dark markings. The skin is smooth, and dorsolateral ridges are absent. The hind feet are webbed, but not extensively.

WHAT DO THE TADPOLES LOOK LIKE? The body can vary from grayish green to dark brown with small dark spots. The tail fin has dark markings, and two thin rows of dark spots or stripes may be present on the tail. Total length is typically less than 3.5 inches.

SIMILAR SPECIES Carpenter frogs are easily distinguished from most other frogs within their geographic range by the four light longitudinal lines on a brown body and the absence of dorsolateral ridges. Juvenile pig frogs look somewhat like adult carpenter frogs but have more webbing on the hind feet.

DISTRIBUTION AND HABITAT In the Southeast, carpenter frogs are found in the Coastal Plain from the Okefenokee Swamp on the northeastern edge of Florida and southeastern Georgia to southeastern Virginia and in an area in northeastern Virginia. Populations are often small and isolated from one another. Carpenter frogs are sometimes called sphagnum frogs because of their strong association with wetlands bordered by sphagnum moss. They are known from swamp ponds, Carolina bays, and open marsh habitat.

BEHAVIOR AND ACTIVITY These highly aquatic frogs seldom venture far from their wetland homes, although they have been observed moving from an aquatic breeding site to other aquatic areas during winter. They spend the winter underwater, presumably buried in the sediments. The males maintain, and aggressively defend, small territories ranging in size from a few square feet up to about 50 square yards.

The call of the carpenter frog sounds like a hammer striking wood.

FOOD AND FEEDING Although the diet has not been documented, it is presumed to include insects, crayfish, and other small invertebrates.

DESCRIPTION OF CALL As the common name implies, the advertisement call of chorusing carpenter frogs sounds something like hammering. Each individual typically repeats his rapid two-note call four or five times. Often only a few individuals may be calling at any time, but when large choruses occur, they sound like a group of carpenters in the distance hurriedly nailing together a house.

REPRODUCTION Carpenter frogs breed from at least May to July in most areas and possibly from April to early fall in warmer regions. Males may

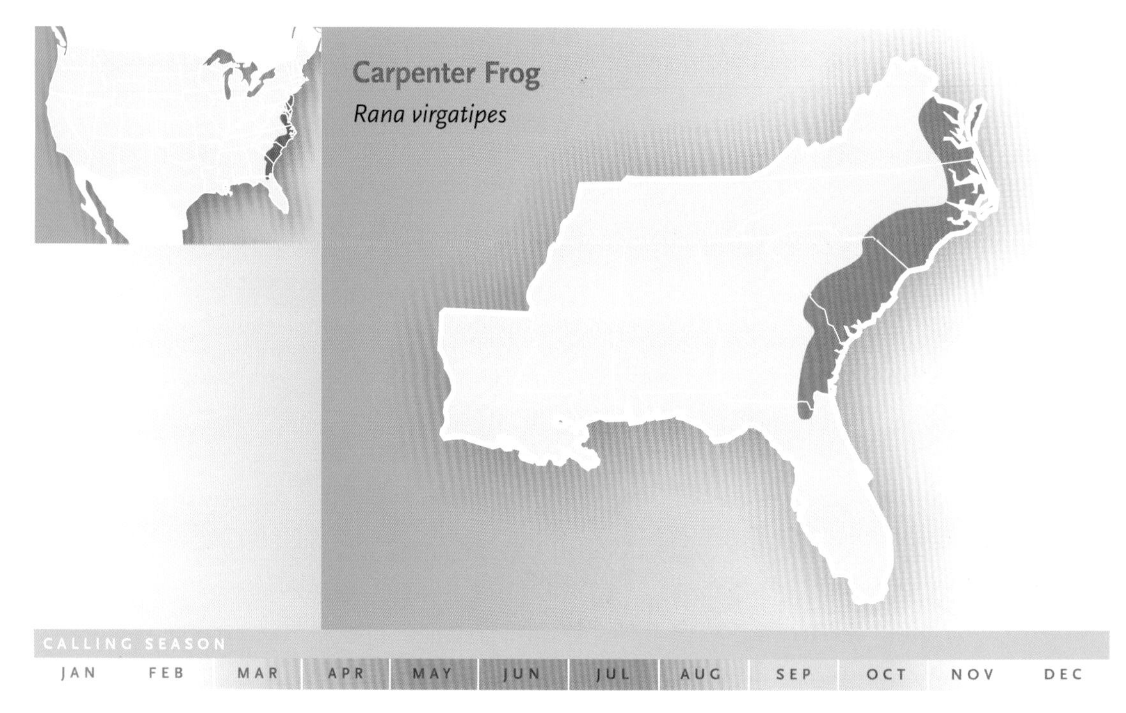

A carpenter frog from the sandhills of North Carolina

begin calling as early as March. Each female lays 200–600 eggs that form small, globular masses. The egg masses may be 2–4 inches across and float below the water's surface, usually attached to vegetation. The water in the breeding ponds is typically more acidic than the water where other frogs breed successfully. The tadpoles remain in the aquatic habitat for a full year and generally emerge in late summer at a body length of about an inch.

PREDATORS AND DEFENSE Watersnakes have been reported to eat carpenter frogs, but these small frogs undoubtedly fall prey to other aquatic predators, including cottonmouths, herons and egrets, large fish, and possibly even large pig frogs, which often share the same habitat. Their primary defense is to remain in the water in thickly vegetated areas or to retreat into the water from the shoreline.

CONSERVATION Conservation needs for carpenter frogs are difficult to assess because so little is known about them. In Virginia, where confirmed populations are uncommon, carpenter frogs are listed as a Species of Special Concern. Carpenter frogs are probably vulnerable in other parts of their range as well due to their somewhat restrictive habitat requirements and the fact that they often occur in small populations.

COMMENTS In 2006, amphibian biologists placed this species in the genus *Lithobates*.

Florida bog frogs have well-defined dorsolateral ridges.

How do you identify a Florida bog frog?

SKIN
Smooth

LEGS
Relatively stocky

FEET AND TOES
Hind feet with moderate webbing

BODY PATTERN AND COLOR
Green to yellowish brown with no markings

DISTINCTIVE CHARACTERS
Small size; no spotting on back; upper lip yellow; dorsolateral ridges present; reduced webbing on hind feet

CALL
Rapidly repeated *chuck-chuck-chuck*

SIZE
max tadpole = 2.25"
typical adult = 1.5"

Florida Bog Frog

Rana okaloosae

DESCRIPTION Florida bog frogs are the smallest true frogs in the Southeast; adults are generally 1.5 to less than 2 inches long. The body is green to yellowish brown and lacks distinctive spotting. Paired dorsolateral ridges extend down most of the back. The belly is light with dark markings. The hind feet are webbed, but less so than those of most other true frogs of the Southeast. The male's paired vocal sacs are below the throat but may not be apparent externally during calling.

A tadpole of the Florida bog frog

WHAT DO THE TADPOLES LOOK LIKE? Florida bog frog tadpoles are greenish brown with yellowish spotting on the tail and white spots on the belly.

SIMILAR SPECIES Florida bog frogs are easily distinguished from other frogs in their geographic range by the combination of dorsolateral ridges, no spotting on the back, and limited webbing on the hind feet. Adults are also much smaller than adults of other true frogs.

DISTRIBUTION AND HABITAT Florida bog frogs have one of the smallest geographic ranges of any frog species in the Southeast. They are known

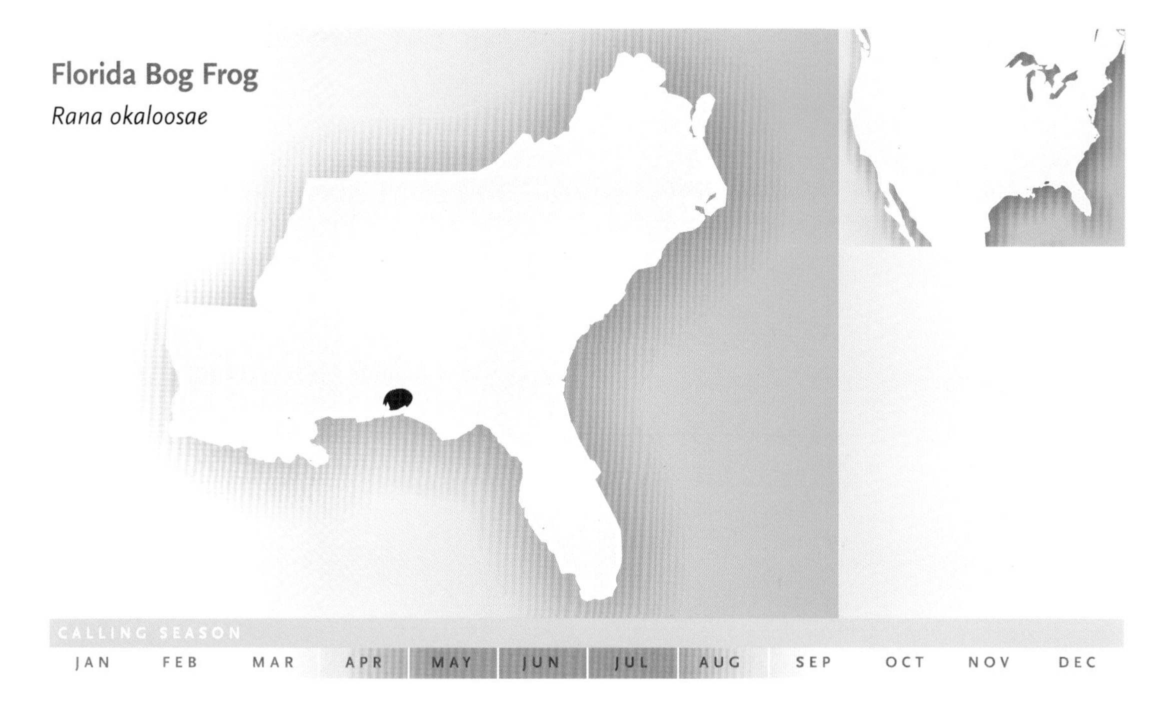

from fewer than three dozen sites in three adjacent counties (Walton, Okaloosa, and Santa Rosa) in the Florida panhandle. Most of the known sites are on Eglin Air Force Base. Juveniles and adults live in shallow, boggy seepage areas associated with small streams that are highly acidic (pH less than 5.5). Sphagnum moss is usually present in the habitat.

A Florida bog frog from Eglin Air Force Base

BEHAVIOR AND ACTIVITY Little is known of the habits of Florida bog frogs. They do not move long distances during the breeding season and are not known to stray far from the seepage habitats where they are typically found.

FOOD AND FEEDING Typical prey probably includes insects and spiders associated with the habitat.

DESCRIPTION OF CALL The advertisement call has been described as guttural *chucks* that are repeated rapidly for 1–4 seconds. Males occasionally give soft single notes that may orient females for breeding purposes.

REPRODUCTION Florida bog frogs breed from April to August. Each female lays a few hundred or more eggs on the surface of standing water; the exact number of eggs is unknown. The tadpoles spend the winter in the aquatic

A Florida bog frog with recently laid eggs

The Florida bog frog has one of the smallest geographic ranges of any native North American frog.

habitat, and froglets emerge during the following spring or summer at a length of about 0.75 inch.

PREDATORS AND DEFENSE Southern banded watersnakes and cottonmouths are presumed to be predators of adults, which may also be eaten by wading birds and carnivorous mammals such as raccoons. Aside from camouflage, the Florida bog frog's primary means of defense is to hop or swim away.

CONSERVATION Florida bog frogs were not discovered until 1982 and are known from only three counties in Florida. Critical conservation measures must include protection of the small streams and surrounding vegetation where these rare frogs are known to occur. Ensuring that known populations remain intact through habitat protection is probably the most effective conservation strategy, and this is an achievable goal because most of the populations are on federal property.

COMMENTS In 2006, amphibian biologists placed this species in the genus *Lithobates*.

Because of their large size and wide mouths, bullfrogs are known predators of many reptiles, birds, and small mammals.

Bullfrog

Rana catesbeiana

DESCRIPTION Bullfrogs are the largest true frogs native to North America. These long-legged, robust frogs are usually green or olive on the back, often with dark mottling that ranges from brown to black. The belly is white to cream but is often mottled with dark pigment. The skin is smooth but may have small bumps scattered across its surface. The rear feet are extensively webbed. The dorsolateral ridges are absent behind the tympanums, and the eyes are relatively large and prominent. Males have paired vocal sacs that extend out from each side of the throat when they are calling. The tympanums of female bullfrogs are about the size of their eyes; those of adult males are twice as large as the eyes.

WHAT DO THE TADPOLES LOOK LIKE? Bullfrog tadpoles are olive green and very large. Scattered black dots on relatively large tadpoles distinguish bullfrog tadpoles from other species. The belly is off-white or cream, and the intestinal coil is generally not visible through the skin. The largest tadpoles reach a total length of about 6 inches.

SIMILAR SPECIES Young bullfrogs are sometimes confused with green frogs, but the latter have a dorsolateral ridge that extends at least part of the way down each side of the body. River frogs are darker brown and have white spots on their lower lip and dark spots on their upper lip. Pig frogs

How do you identify a bullfrog?

SKIN
Smooth

LEGS
Very muscular and long

FEET AND TOES
Hind feet almost fully webbed

BODY PATTERN AND COLOR
Back generally olive green, often with dark brown mottling

DISTINCTIVE CHARACTERS
Large size; no dorsolateral ridges

CALL
Loud, low-pitched *jug-o-rum.*

SIZE
max tadpole = 6.5"
typical adult = 5

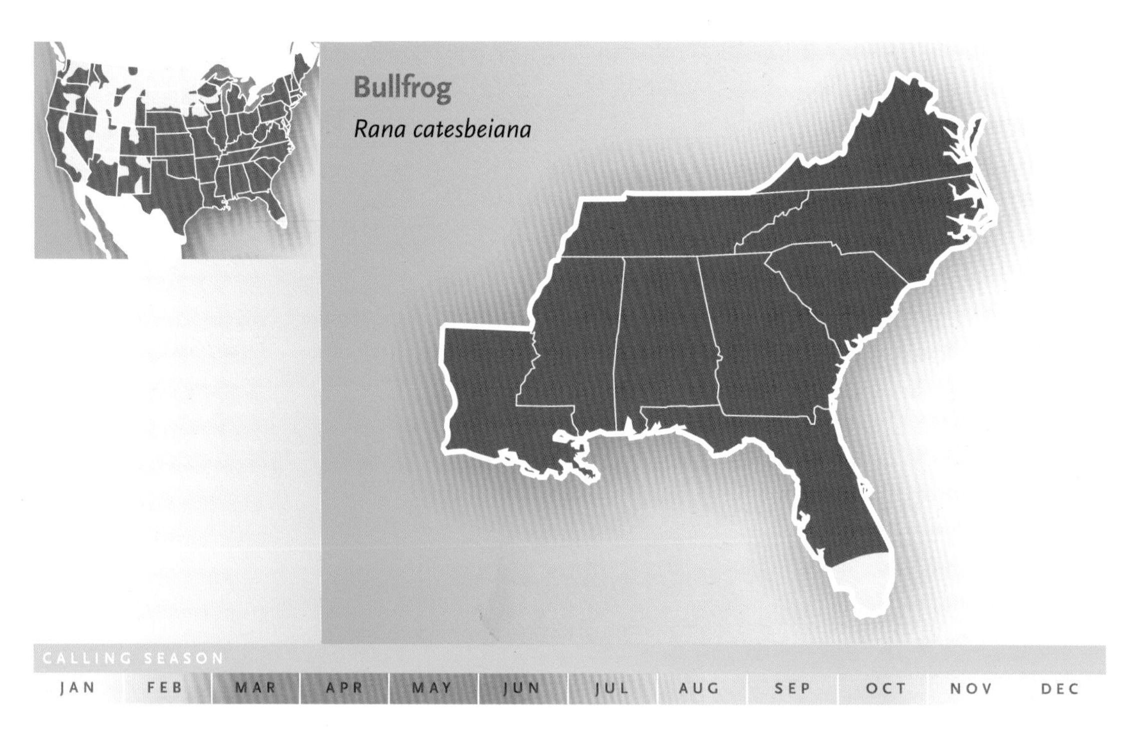

Bullfrogs require permanent aquatic habitats, within which their movements are usually confined to a very small area.

look a lot like bullfrogs, but they have a more pointed snout and the webbing on their hind feet extends all the way to the end of the longest toe.

DISTRIBUTION AND HABITAT Bullfrogs are found in a variety of aquatic habitats throughout the eastern United States and are present throughout most of the Southeast except for the southern half of the Florida peninsula and most of the Okefenokee Swamp. In the western United States and elsewhere, they are an invasive species and often cause major conservation problems. Bullfrogs can be found in nearly any permanent aquatic habitat, including man-made ponds, lakes, and reservoirs as well as sluggish backwaters of rivers, ditches, and swamps. They generally prefer warmer water bodies. Juveniles are adept at colonizing newly created ponds and often disperse relatively long distances overland.

BEHAVIOR AND ACTIVITY Bullfrogs are primarily nocturnal, although they sometimes bask in the sun. They can typically be found sitting at the edge of a water body or in shallow water. When disturbed, they often jump quickly into the water. They usually hibernate underwater, buried in the mud, but some individuals overwinter in burrows on land. They emerge from hibernation later in the spring than most other frogs. Their movements are usually confined to a very small area within their aquatic home, although some make overland journeys to find new aquatic habitat. The tadpoles are solitary and move into deeper water just before metamorphosis. When held in the hand, bullfrogs often go completely limp.

FOOD AND FEEDING Bullfrogs eat virtually anything they can swallow, and that includes most other North American frogs. Some remarkable prey items that have been documented include leeches, centipedes, scorpions, fish, other frogs, small alligators, turtles, snakes, rodents, bats, birds, and even a young mink. Bullfrogs hunt primarily by ambush, waiting for an animal smaller than they are to approach. They may even orient toward distress calls of other frogs or movement in the water. In dense populations, large bullfrogs can be highly cannibalistic on smaller ones.

DESCRIPTION OF CALL The advertisement call is a loud bass note generally described as sounding like *jug-o-rum* or sometimes *more rummmm* or *brrruuuumm*; it is usually given while the frog is sitting in shallow water. Call transmission underwater as well as in the air may be important for attracting females and dissuading male competitors. When approached by another bullfrog, male bullfrogs produce a territorial call described as *phoot*. The distress call—a loud, high-pitched scream—is given with the mouth open when a bullfrog is being attacked. Some scientists

Did you know?

Although males of many species of frogs engage in physical combat with each other, the bullfrog is the only frog in the Southeast with adult males larger than adult females.

Most bullfrogs are mottled with brown spots.

Bullfrogs are very aquatic but can sometimes be found moving overland.

believe the purpose of this scream is to attract mammal or bird predators that may interrupt the attack and give the bullfrog a chance to escape.

REPRODUCTION In most parts of the Southeast, bullfrogs breed from March to October. They are very territorial, and males will wrestle with each other to establish dominance. Male bullfrogs mature at a smaller size and earlier age (1–2 years after metamorphosis) than females (2–3 years after metamorphosis). Females select mates by choosing the territory of a male. Older females tend to be pickier than younger females when choosing a mate, and have even been known to vocalize like males to increase the number of males calling, thus maximizing the pool of potential mates. Young "satellite males" that do not yet have territories may lurk at the edges of a larger male's territory and intercept females. Amplexus and egg-laying generally occur in vegetation-choked areas of shallow permanent water. Larger, older females may produce two clutches per year, especially in warm regions, and can lay up to 20,000 eggs per clutch. Eggs are laid in thin sheets on the water's surface and may cover an area 3 feet square. The eggs hatch 3–5 days after they are laid, and the tadpole stage lasts from a few months to more than 2 years.

PREDATORS AND DEFENSE Major predators of bullfrogs include humans, who consume their legs. Alligators, snakes, large wading birds, and mammals such as raccoons eat large adults. Natural predators on eggs include leeches, fish, and salamanders. Many predators find the eggs and tadpoles unpalatable, though, allowing bullfrogs to persist in habitats containing fish. Some researchers have suggested that adult bullfrogs are partially resistant to the venom of cottonmouths and copperheads. The *squak* bullfrogs give when fleeing may alert other frogs to potential danger. When in the grasp of a predator such as a watersnake, a bullfrog will emit a loud, piercing scream. Despite their vulnerability to natural predators at early life stages and smaller sizes, bullfrogs have been known to live in the wild for up to 8–10 years, and some individuals may live as long as 15 years.

CONSERVATION Bullfrogs require permanent aquatic habitats, and elimination or extensive pollution of these will reduce their populations. In the Southeast, however, bullfrogs continue to be prevalent in most water bodies. Some naturally occurring diseases may become problems for populations stressed by pesticides, pH changes, and other human-caused pollution. Although bullfrogs are important components of aquatic ecosystems in the Southeast, they are not native to the western United States, where they have been widely introduced and are now a major threat to many species of native frogs. Permanent ponds, such as those found on golf courses and farm ponds, in the western United States provide artificial habitat for bullfrogs that would otherwise be unavailable. Bullfrogs have also been introduced into many other countries, including Mexico, Cuba, Jamaica, Japan, the Republic of Korea, and many parts of Europe and South America.

The large tympanum of this bullfrog distinguishes it as a male.

COMMENTS In 2006, amphibian biologists placed this species in the genus *Lithobates* and now refer to it as *Lithobates catesbeianus*.

Did you know?

The tympanum (external eardrum) of a male bullfrog is larger than that of the female, suggesting that it is as important for males to hear advertisement calls as it is for females. Research suggests that the sound of the bullfrog's call is partially transmitted out through the tympanums in addition to the vocal sacs.

A pig frog from Louisiana

How do you identify a pig frog?

SKIN
Smooth

LEGS
Very muscular and long

FEET AND TOES
Complete webbing between hind toes

BODY PATTERN AND COLOR
Green to dark brown with dark markings on the lower back and sides

DISTINCTIVE CHARACTERS
No dorsolateral ridges; extensive webbing of hind feet

CALL
Piglike grunts

SIZE
max tadpole = 4"
typical adult = 5"

Pig Frog

Rana grylio

DESCRIPTION The largest pig frogs are only slightly smaller than bullfrogs, reaching body lengths of 6.5 inches. The hind legs are sturdy, and the snout is somewhat pointed. The toes on the hind feet are almost completely webbed, and no dorsolateral ridges are present behind the tympanum. The color of the back is variable, ranging from green to olive to dark brown with dark flecks or spots, especially toward the hind end. The belly is usually solid white but may be grayish or have dark grayish patterns. Juvenile pig frogs have four stripes—a pair running lengthwise down each side of the back in a pattern similar to that of carpenter frogs.

WHAT DO THE TADPOLES LOOK LIKE? Pig frog tadpoles can be greenish, golden brown, dark brown, or black with or without light yellow spotting on the body. The top portion of the tail fin is translucent and characteristically bears a distinctive row of small black spots that form a line extending the length of the fin. The belly is white with dark mottling, but the chin is often dark. The tadpoles are large, reaching a total length of about 4 inches; froglets have a body length of 1.5–2 inches at metamorphosis.

A pig frog tadpole

SIMILAR SPECIES Pig frogs are most often confused with bullfrogs, although some individuals may resemble other true frogs. The lack of dorsolateral ridges separates pig frogs from most other true frogs; the extensive webbing on the hind feet, which extends to the tip of the longest (fourth) toe, and more pointed snout distinguish the pig frog from the bullfrog, carpenter frog, and river frog.

DISTRIBUTION AND HABITAT Although pig frogs have been successfully introduced into Puerto Rico and the Bahamas, the natural geographic range of the species is limited to the Coastal Plain from central South Carolina throughout Florida and across Louisiana to eastern Texas. The ideal habitat typically features permanent water, especially large lakes or reservoir coves with lily pads and abundant emergent vegetation, flooded rice fields and ditches, freshwater marshes, and swamp pools. Populations can also persist in large, heavily vegetated Carolina bays that dry periodically. Pig frogs are reported to be among the more successful frogs at colonizing urban or suburban lakes in some parts of their geographic range.

BEHAVIOR AND ACTIVITY Pig frogs are noted for being very aquatic, and adults usually remain in permanent wetlands. In aquatic habitats that occasionally dry, the frogs will burrow into the mud or organic layer of the wetland until it fills again. Individuals may also move between wetlands when conditions are unsuitable. Pig frogs are active primarily at night but often call during the day. Like bullfrogs, males are quite territorial during the breeding season and do not tolerate intrusions by other males. In the

The webbing on a pig frog extends to the tip of the longest toe on the hind foot, distinguishing it from the very similar bullfrog.

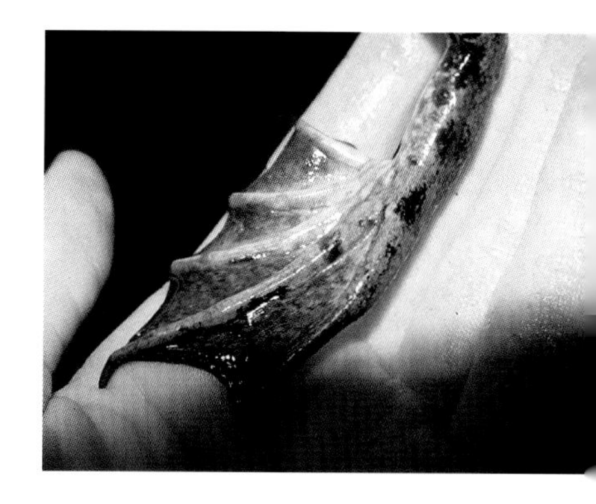

Pig Frog
Rana grylio

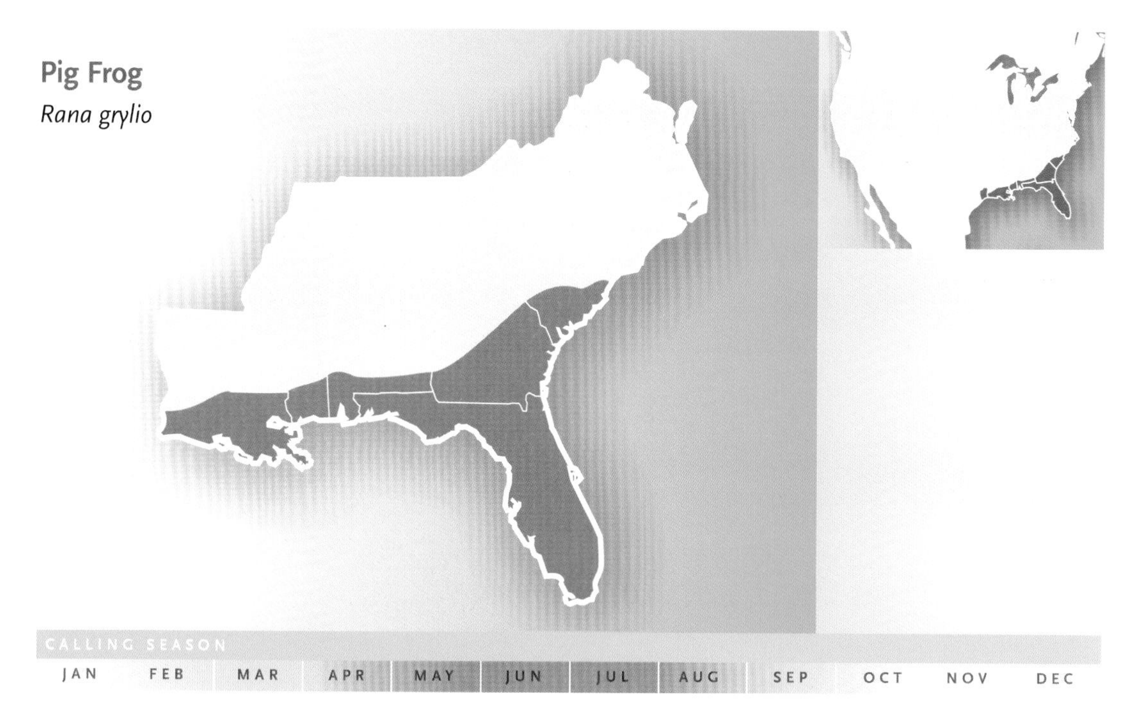

Pig frogs are never found far from water.

cooler parts of the geographic range, pig frogs hibernate, presumably burying themselves in the mud at the bottom of their pond, from as early as November to as late as March.

FOOD AND FEEDING Adults eat crayfish; a wide variety of insects, especially beetles and dragonflies; leeches; other frogs, including smaller pig frogs; and even small fish and snakes. Tadpoles presumably eat primarily algae and plant detritus.

DESCRIPTION OF CALL Its advertisement call is responsible for the pig frog's common name. A single individual utters several hollow-sounding grunts at a time; a full chorus produces an uninterrupted roar. Many people mistake calling pig frogs for alligators, not because pig frog calls sound anything like an alligator, but because people see alligators and then hear pig frogs, which are often in the same habitats.

REPRODUCTION Pig frogs breed during most of the year in the southern parts of their geographic range, and males call at least from February to September in most areas. The open marsh, lake, and swamp habitats where they breed may not be the sites where they spend the rest of the year; some individuals migrate to breeding areas. A female can lay from 8,000 to as many as 15,000 eggs, depositing them in masses that float on the surface, usually supported by vegetation. The eggs typically hatch into tadpoles in 2–3 days. Pig frogs sometimes remain in the tadpole stage for more than a year and possibly longer before undergoing metamorphosis. Males establish territories during the breeding season and will engage in physical combat with each other, actually wrestling for several minutes until one is defeated and leaves the area.

Did you know?

Frog legs really do taste like chicken to most people.

PREDATORS AND DEFENSE Wading birds, ospreys, large watersnakes, and cottonmouths are known predators. Given pig frogs' strong aquatic tendencies, other predators probably include alligators and large fish.

CONSERVATION Pig frogs are common in most parts of their geographic range and do not appear to be in need of protection in any state. At the edges of their range, where they tend to be less common, local populations may be vulnerable to extirpation as the result of human activities such as development and wetland alteration.

COMMENTS In 2006, amphibian biologists placed this species in the genus *Lithobates*.

River frogs may be extinct in North Carolina.

River Frog

Rana heckscheri

DESCRIPTION River frogs are among the largest frogs in North America; large adults can be as big as an average-sized bullfrog and have powerful back legs. The body is dark green, brown, or black above; the underside is characteristically dark with lighter markings, often including a white crescent that runs from the front of each hind leg to the groin. The typical color patterns prevail in most specimens, but the green, brown, or black coloration on the back and the dark and light shading on the belly can be highly variable, making precise reliable description difficult. Although similar in general appearance to bullfrogs and pig frogs, river frogs' skin is often rougher. The hind feet are extensively webbed, but the web on the fourth (longest) toe does not extend to the tip. River frogs do not have dorsolateral ridges. The tympanum is large, and the eyes are prominent, with juveniles often retaining the red eye coloration of the tadpoles.

WHAT DO THE TADPOLES LOOK LIKE? In overall appearance, river frog tadpoles are among the most impressive tadpoles in North America because of

River frog tadpoles are easily recognizable because of their large size and red eyes.

How do you identify a river frog?

SKIN
Smooth

LEGS
Long and muscular

FEET AND TOES
Toes pointed with extensive webbing

BODY PATTERN AND COLOR
Back greenish brown to black; belly dark gray with white markings

DISTINCTIVE CHARACTERS
Large size; no dorsolateral ridges; lips dark with white spots

CALL
Long, drawn-out snore with occasional snorts and grunts

SIZE
max tadpole = 6.25"
typical adult = 4"

River frog tadpoles often travel in large schools.

their large size (more than 6 inches long), striking coloration (mostly black on the body with a lighter-colored tail bordered in black), and distinctive bright red eyes. They are often found in large schools of hundreds of individuals that move in unison and concentrate in shallower water during the day.

SIMILAR SPECIES River frogs are most difficult to distinguish from bullfrogs and pig frogs, although the snorelike advertisement call readily separates them from the latter two species. River frogs usually have a darker belly, and the lips, especially the lower lip, have white spotting. The lack of dorsolateral ridges distinguishes the river frog from leopard frogs, green frogs, and gopher frogs at all sizes.

DISTRIBUTION AND HABITAT The river frog is restricted to the Southeast, historically being found sporadically from southern North Carolina to the northern third of Florida and across the Florida panhandle to southeastern Mississippi. These frogs can be abundant in local but widely separated populations; they have not been observed or captured in North Carolina since the mid-1970s and may be extinct in that state. River frogs are typically associated with swampy habitats contiguous with streams, lakes, oxbows, ponds, and rivers with heavy aquatic vegetation, although they are not characteristic of the diverse range of wetlands that leopard frogs, green frogs, and bullfrogs inhabit. Adult river frogs are not known to make seasonal migrations or to move long distances overland, but young individuals apparently disperse from the aquatic areas where they developed.

BEHAVIOR AND ACTIVITY Adults and juveniles are almost exclusively nocturnal. They are considered rare in most regions but may be common in localities where habitats of both tadpoles and adults are protected. River frogs are usually not as wary as other large frogs, and can often be easily observed and captured by hand at night. Once captured, they are also easy to handle and less likely to jump than leopard frogs or green frogs. River frog tadpoles are noted for congregating in large schools that may look like a large fish or alligator as they move through the water.

FOOD AND FEEDING Although direct observations of the diet are limited, river frogs probably eat most small animals they encounter, especially insects and smaller frogs.

DESCRIPTION OF CALL The advertisement call has been described as snoring punctuated with snorts, snarls, grunts, and other guttural sounds. Their distinctive call is perhaps the most reliable means of distinguishing river frogs from bullfrogs and pig frogs in natural situations. Some people compare the river frog call with the bellowing of a distant alligator.

REPRODUCTION River frog males begin calling in the spring (usually April), and breeding continues sporadically until August depending on environmental conditions that are not readily predictable or completely understood. Mating occurs in aquatic areas, with the eggs being laid in still waters of swamps or ponds associated with rivers and streams. The territorial aggression among males characteristic of bullfrogs has not been observed in river frogs. Adult males (ca. 3–5 inches) are typically smaller than mature females (ca. 4–6 inches). A single female may lay from 1,000 to as many as 8,000 eggs in a single layer on the water's surface. Eggs have been reported to develop into larvae in as few as 3

River frogs vary in color from light brown to dark brown.

River Frog

Rana heckscheri

A young river frog from South Carolina

A river frog transforming from a tadpole into an adult

and as many as 15 days. The larvae sometimes remain in the aquatic habitat as tadpoles for up to 2 years, passing through one or two winters. Froglets are slightly over 1 inch to almost 2 inches in length at metamorphosis.

PREDATORS AND DEFENSE Little is known about river frog predators. Southern banded watersnakes and probably large fish such as bass prey on tadpoles, and grackles are reported to eat recently metamorphosed young. Speculation that toxic skin secretions make river frogs unpalatable to some predators remains to be proved, but that trait could account for the limited number of documented predation records.

CONSERVATION Due possibly to their patchy distribution and their rarity in many of the places where they do occur, few population studies have been conducted on river frogs, making it difficult to assess their status throughout most of their geographic range. The species remains common, if not abundant, on protected properties in southwestern South Carolina. However, their disappearance from North Carolina during the latter part of the twentieth century and their dependence on swamp systems that may be degraded by development and construction activities throughout the Southeast are indications that remaining populations should be considered in future conservation planning. In areas where highways have been constructed near river frog populations, the sounds of nighttime traffic will undoubtedly mask or interrupt the snorelike breeding call of males.

COMMENTS In 2006, amphibian biologists placed this species in the genus *Lithobates.*

The crawfish frog's white belly, underside of the chin, and throat do not have dark markings.

Crawfish Frog

Rana areolata

DESCRIPTION Crawfish frogs are much stockier than other true frogs within their geographic range, although adults are usually less than 3.5 inches in length. The legs are relatively stout and short, and the skin is smooth. The coloring above is variable but typically consists of dark spots on a lighter gray or tan background, sometimes with each spot appearing to be encircled by a lighter ring. The groin area and back of the thighs have a yellowish wash; the white belly, underside of the chin, and throat do not have dark markings. The hind feet have slight webbing that does not extend far up from the base of the toes. The tympanum is about the same size as the eye. Mature females are larger than males.

WHAT DO THE TADPOLES LOOK LIKE? Crawfish frog tadpoles resemble the tadpoles of some of the other true frogs with which they occur, such as leopard frogs and pickerel frogs, but can be distinguished by careful examination of the teeth in the upper jaw; the inner row of teeth is separated into two halves by a gap. They are slightly more than an inch long prior to metamorphosis.

SIMILAR SPECIES Crawfish frogs can be distinguished from most other frogs within their geographic range by their stout, stocky build (leopard frogs and pickerel frogs are much more slender, and both have distinct

How do you identify a crawfish frog?

SKIN
Mostly smooth

LEGS
Stout and relatively short

FEET AND TOES
Limited webbing between toes

BODY PATTERN AND COLOR
Upper surface of legs and back with large, dark spots; belly completely white; yellow groin area

DISTINCTIVE CHARACTERS
Stocky, robust body; relatively short legs; distinct dorsolateral ridges; solid white belly

CALL
Continuous guttural snore

SIZE
max tadpole = 2.5"
typical adult = 2.75"

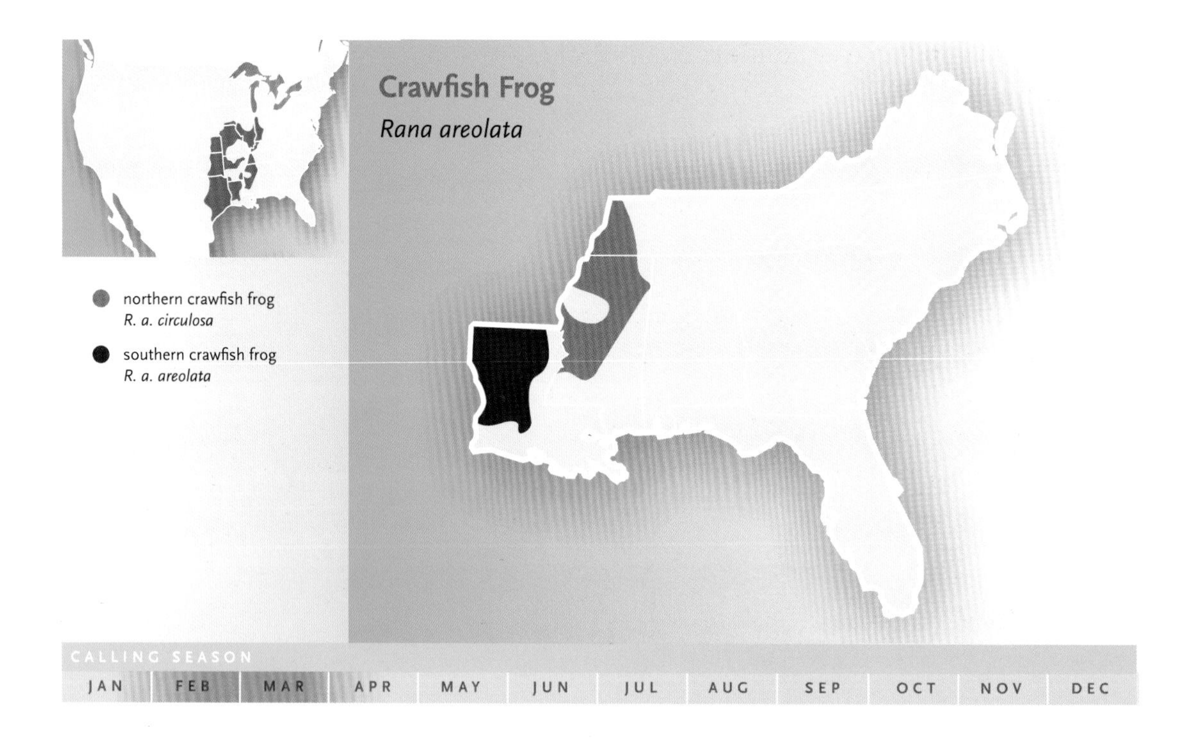

spots on the back); the presence of complete, prominent dorsolateral ridges (absent in bullfrogs and only partially present in green frogs); and the pure white belly. Their low-pitched, snorelike advertisement call should not be confused with any other species of frog except possibly pickerel frogs.

DISTRIBUTION AND HABITAT The spotty distribution pattern includes some midwestern states as well as the states from Missouri south to eastern Texas and west to Indiana and south to Mississippi. In the Southeast, the species is intermittently distributed through much of the northern two-thirds of Mississippi, western Tennessee, and in scattered localities in northern and western Louisiana. Two subspecies are recognized: the southern crawfish frog (*R. a. areolata*) is found in Louisiana, and the northern crawfish frog (*R. a. circulosa*) is found in Mississippi and north through western Tennessee to Kentucky. Adults breed in seasonal isolated wetlands or temporary aquatic sites but spend most of their life on land in habitats such as stream and river floodplains, mixed hardwood and pine forests, and open pasturelands. They spend much of their time underground, often taking refuge in crayfish burrows or tunnels dug by small mammals and beneath logs or other natural ground litter. Adults move overland from underground burrows to aquatic breeding sites, but these are often only a few yards away.

BEHAVIOR AND ACTIVITY A distinctive behavioral feature of crawfish frogs is their use of crayfish burrows or other holes in the ground—some extending as much as 5 feet beneath the surface—as refuges and hibernation sites. Individuals sometimes sit on a flattened area at the entrance to the burrow. They are active primarily during the early evening and at night. Both sexes can live to be at least 5 years old.

FOOD AND FEEDING Tadpoles eat algae and microscopic plants in the water column. Adults eat a wide variety of insects such as ants, beetles, and crickets, and other invertebrates including crayfish and spiders. They probably eat larger prey such as small mammals or smaller frogs when the opportunity presents itself.

DESCRIPTION OF CALL The advertisement call is a low-pitched, guttural snore that can be heard as much as half a mile away. A large chorus may sound like a group of hogs snorting.

REPRODUCTION Crawfish frogs in the southeastern part of their range may begin calling from seasonal wetland habitats on warm nights (above 50° F) after migrat-

The northern subspecies of the crawfish frog

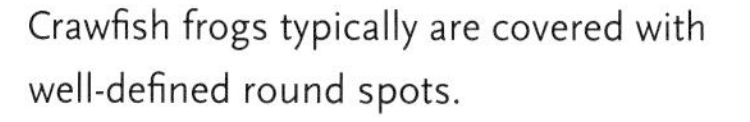

Crawfish frogs typically are covered with well-defined round spots.

ing during or after steady rains. Males sometimes arrive at the wetland several days before females. Because these frogs are primarily explosive breeders, chorusing peaks during the first few days and then dwindles over several days or weeks. The choruses are relatively small, often consisting of only a few individuals. Eggs are laid on the surface or on vegetation in shallow water, and several females may deposit their eggs in a cluster, with each laying from 2,000 to as many as 7,000 relatively large eggs (diameter up to 0.1 inch). Larvae hatch from the eggs within 1–2 weeks and may take 2–3 months to metamorphose. Recent metamorphs are slightly more than an inch long.

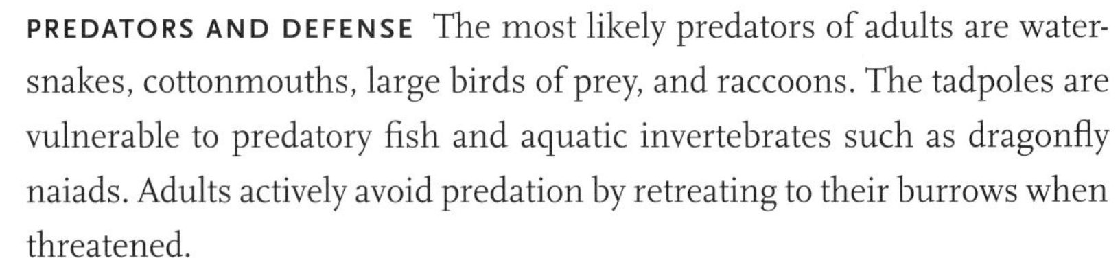

PREDATORS AND DEFENSE The most likely predators of adults are watersnakes, cottonmouths, large birds of prey, and raccoons. The tadpoles are vulnerable to predatory fish and aquatic invertebrates such as dragonfly naiads. Adults actively avoid predation by retreating to their burrows when threatened.

Crawfish frogs are stockier than other true frogs within their geographic range.

CONSERVATION Crawfish frogs appear to be uncommon in most areas of their range, although their actual status is often difficult to determine because they spend so much time underground. Certain land use practices associated with agriculture, road construction, and urban development (e.g., paving, mowing, and soil disturbance) may harm localized populations by destroying burrowing habitat. These frogs are also vulnerable to road mortality when highways lie between their terrestrial burrows and wetland breeding sites.

COMMENTS In 2006, amphibian biologists placed this species in the genus *Lithobates* and now refer to it as *Lithobates areolatus*.

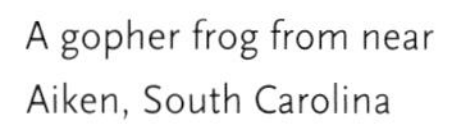

A gopher frog from near Aiken, South Carolina

Gopher Frog

Rana capito

DESCRIPTION Gopher frogs are characterized by their stout body and legs. Adults are generally 3–3.5 inches long. The coloring above is highly variable both geographically and among individuals, with those in much of Florida having a lighter background that can be cream colored, light gray, or brownish yellow. The upper surface of those from other areas is usually darker, and some specimens even appear uniformly dark at times. Dark blotches are present on the back and upper parts of the legs, but the contrast is greater on individuals with a lighter background color. Webbing on the hind feet is minimal. The chin and belly have dark mottling. The raised dorsolateral ridges may be yellowish.

WHAT DO THE TADPOLES LOOK LIKE? Gopher frog tadpoles have a greenish brown, unspotted body and a long tail that ranges from clear to spotted. They are most similar in appearance to leopard frog tadpoles. The tadpoles are 1–1.5 inches long when they hatch and can reach a total length of 3.5 inches prior to metamorphosis.

SIMILAR SPECIES The heavy body; short, stout legs; thick dorsolateral ridges; and yellow on the upper lip and underside will distinguish gopher frogs from most other true frogs with which they occur naturally. Their low-pitched and continuous snoring advertisement call is distinctive as well.

How do you identify a gopher frog?

SKIN
Smooth but with large warts

LEGS
Relatively short and robust

FEET AND TOES
Limited webbing

BODY PATTERN AND COLOR
Light or dark brown with large, dark spots; yellow on upper lip, under arms, and in groin area

DISTINCTIVE CHARACTERS
Heavy body with short, thick limbs; dorsolateral ridges prominent; belly with dark markings

CALL
Long snore

SIZE
max tadpole = 3.5"
typical adult = 3"

DISTRIBUTION AND HABITAT Gopher frogs occur in scattered localities in the Gulf and Atlantic coastal plains from southern Alabama to east-central North Carolina and southward except for most of the southern tip of Florida. Isolated populations have been discovered in central Alabama and Tennessee. Herpetologists once recognized three subspecies: the Carolina gopher frog (*R. c. capito*), the dusky gopher frog (*R. c. sevosa*), and the Florida gopher frog (*R. c. aesopus*). Subsequent genetic analyses indicated that the dusky gopher frog should be considered a full species (*R. sevosa*), and that populations from Florida and other parts of the geographic range do not warrant subspecies status. Adults and juveniles occupy dry habitats such as longleaf pine–scrub oak forests as well as bottomland hardwoods and pine flatwoods. In areas where gopher tortoises occur, the frogs often use tortoise burrows as refuges during periods of drought or extreme winter cold. In other areas they seek refuge in mammal burrows, crayfish holes, or other underground openings or tunnels. Most aquatic breeding sites are seasonal wetlands in both upland and lowland (i.e., flatwoods) habitats that lack predatory fish.

Gopher frogs spend most of the daytime hours underground, commonly in burrows made by other animals. The frogs often create a flattened area in front of the burrow on which they may sit after emerging.

BEHAVIOR AND ACTIVITY Gopher frogs spend most of the daytime hours, as well as dry or cold periods, underground in burrows made by other animals or in stump holes or decayed root holes. Breeding adults may move more than a mile during the night to reach a wetland site where they mate and lay eggs, although some individuals occupy burrows quite close to their breeding site. Gopher frogs often create a flattened area in front of a crayfish, mammal, or tortoise burrow on which they sometimes sit after emerging. Typically, a single individual occupies each burrow, and males may have territories that they defend from other males. Although large local populations are present in some areas, gopher frogs are generally rare, and in some years no individuals are seen over wide portions of the geographic range.

A gopher frog from Florida

Gopher Frog

Rana capito

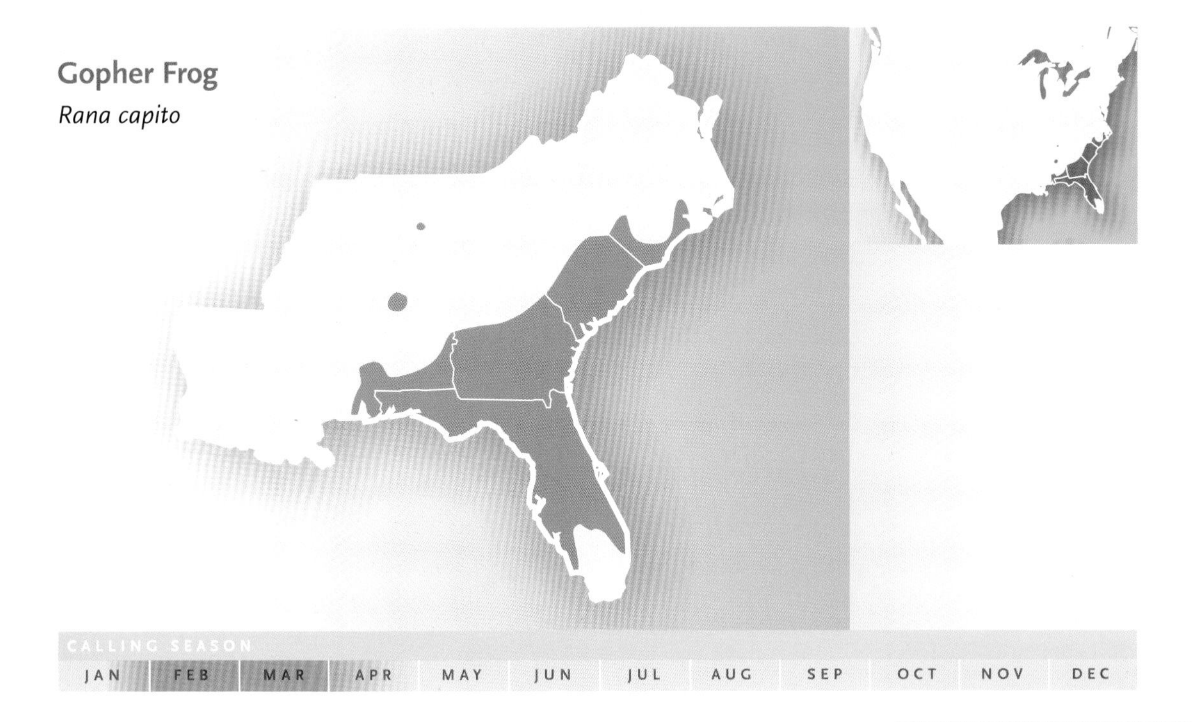

FOOD AND FEEDING Gopher frogs feed primarily at night on insects, spiders, and worms. They have also been known to eat smaller frogs and toads. The exceptionally large mouth allows individuals to swallow prey items nearly as large as themselves. Reports indicate that gopher frogs that feed on adult toads are sometimes affected by the toads' toxins, but the frog apparently survives and reaps the benefits of an exceptionally large meal.

DESCRIPTION OF CALL The advertisement call is a deep snore that may continue without interruption for several seconds. A chorus may sound like a low-pitched roar. Males call from within the wetland habitat where breeding occurs and have been reported to call underwater, where the low-pitched sound probably travels well.

REPRODUCTION Gopher frogs breed most commonly in late winter and into the spring, but will breed following heavy rains, such as hurricane-related storms during the fall, even when nighttime temperatures are cool. In the southern parts of the range, adults may call or mate at any time of year. The males usually arrive at the wetlands earlier than females and remain longer. The females lay between 1,000 and 2,000 eggs in a rounded cluster usually attached to or resting on vegetation in shallow water. Because the adults can breed throughout the year if conditions are suitable, the tadpoles may metamorphose in as few as 4 months or as many as 7 months, depending on the water temperature. Gopher frogs can live more

Did you know?

Male gopher frogs call underwater. Low-frequency calls, such as their "snore," carry well underwater, and female gopher frogs can hear them but potential predators along the shore may not.

than 6 or 7 years and presumably remain reproductively active throughout their lives.

PREDATORS AND DEFENSE Southern banded watersnakes are probably important predators of adults and tadpoles. Large birds of prey such as hawks, and probably owls, also eat adults. Aquatic insects (caddis fly larvae) and salamanders (eastern newts) eat gopher frog eggs. Presumably the frogs and tadpoles are prey for turtles, various species of snakes, wading birds, and raccoons as well. Wetland breeding sites are nearly always free of predatory fish. When threatened, an adult gopher frog will lower its head and place its front feet over its eyes.

CONSERVATION Gopher frogs are declining throughout most of their geographic range, and the species is a potential victim of a variety of urban development and agricultural and forestry practices. Their dependence on suitable upland habitats and fish-free breeding sites makes gopher frogs particularly sensitive to environmental disturbance. Their native habitat included forests that historically experienced natural fires, and modern forestry practices that suppress fire can be detrimental to gopher frogs by increasing the acidity of breeding ponds, allowing understory hardwoods to grow, and changing wetlands' hydrology. The loss of gopher tortoise burrows as a consequence of development and the removal of stump holes in some forestry management plans can eliminate underground refuges for gopher frogs, and the loss or modification of seasonal wetlands eliminates their breeding sites. Gopher frogs are officially recognized as Threatened in Alabama and Florida, as Rare in Georgia, and as a Species of Special Concern in North and South Carolina.

A gopher frog seeks shelter under pine needles.

COMMENTS In 2006, amphibian biologists placed this species in the genus *Lithobates*.

The dusky gopher frog is probably the most endangered species of frog in the United States.

How do you identify a dusky gopher frog?

SKIN
Fairly rough

LEGS
Stocky

FEET AND TOES
Hind feet with minimal webbing

BODY PATTERN AND COLOR
Upper body dark gray or brown with large, darker brown spots; belly with dark markings

DISTINCTIVE CHARACTERS
Heavy body with short, thick limbs and large head; prominent dorsolateral ridges

CALL
Steady snore

SIZE
max tadpole = 3"
typical adult = 3"

Dusky Gopher Frog *Rana sevosa*

DESCRIPTION Dusky gopher frogs have a stout body and legs, and the skin on the back has a warty or bumpy appearance. Adults range in length from about 2 to slightly less than 4 inches, and females usually grow larger than males. These frogs are darker than the other gopher frogs. The dark blotches on the back and upper parts of the legs may show little contrast on the darkest individuals. The chest and the underside of the chin have dark spots, and the webbing on the hind feet is not obvious.

WHAT DO THE TADPOLES LOOK LIKE? Dusky gopher frog tadpoles are difficult to distinguish from leopard frog tadpoles. The body is greenish brown and lacks spots.

A tadpole transforming into a young dusky gopher frog

SIMILAR SPECIES Dusky gopher frogs are easily differentiated from other frogs within their small geographic range by the combination of stout body and legs, thick dorsolateral ridges, and overall dark coloration. Their deep snore is a distinctive advertisement call. They are essentially impossible to distinguish from gopher frogs from nearby southern Alabama.

DISTRIBUTION AND HABITAT This is one of the most localized frog species in the United States. It was once known only from the southernmost counties in Mississippi, a few parishes in eastern Louisiana, and a single county in Alabama. As of 2006, only two populations are known to exist, one in Jackson County, Mississippi, and the other in Harrison County, Mississippi. The exact historical range of the species remains poorly known, and not all herpetologists agree that the dusky gopher frog is a species distinct from the other gopher frog species. Adults live in upland habitats, especially favoring longleaf pine forests, and retreat into gopher tortoise burrows or other underground refuges during most of the year. They breed in small, shallow seasonal wetlands. Juveniles are assumed to hide in underground burrows also.

A dusky gopher frog from Mississippi

BEHAVIOR AND ACTIVITY Except when foraging or engaged in breeding activities, dusky gopher frogs are usually underground. They search for prey near their burrow, and many have a platform of compacted soil at the

The skin on the dusky gopher frog's back has a warty or bumpy appearance.

Dusky Gopher Frog

Rana sevosa

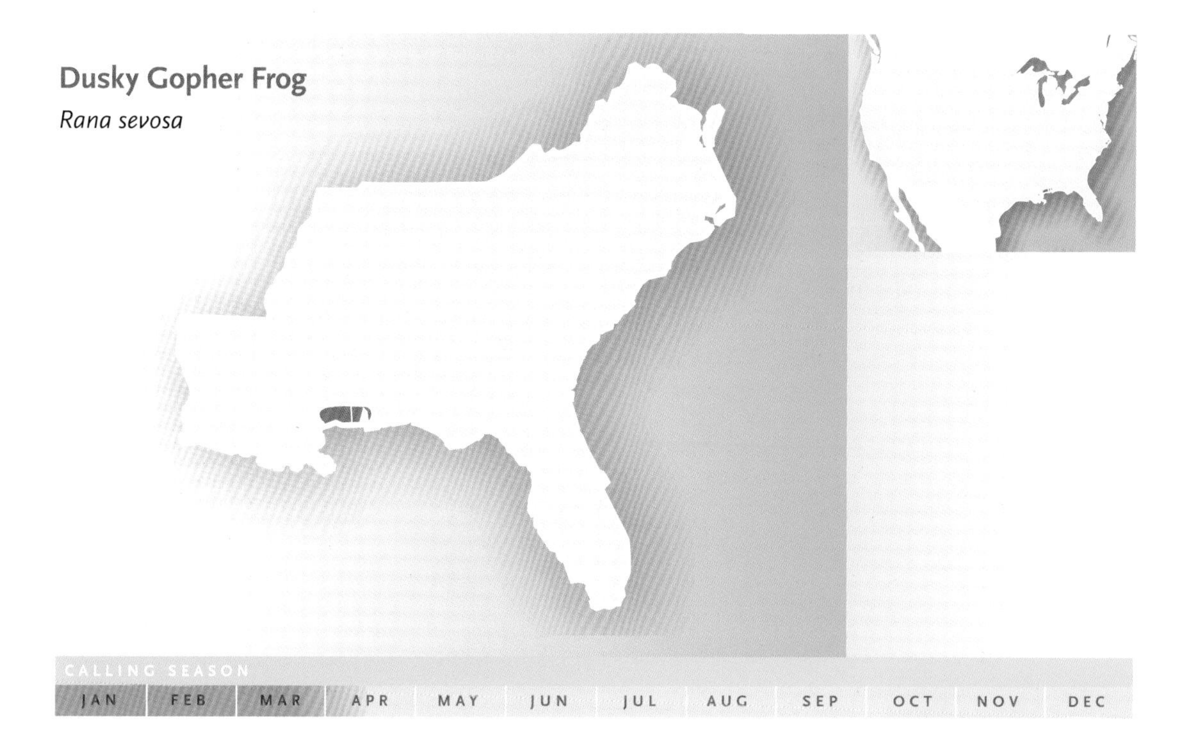

burrow entrance where they sit. During the winter reproductive period, adults may travel overland as much as 300 yards to aquatic breeding sites. Males may set up territories within the wetland that they defend in combat with other males.

FOOD AND FEEDING Dusky gopher frogs are known to eat a variety of beetles but probably also prey on other insects, spiders, and worms. Their exceptionally large mouth presumably allows them to eat rather large prey, including other frog and toad species.

DESCRIPTION OF CALL The advertisement call is a deep, continuous snore. Males are known to call while underwater.

REPRODUCTION Dusky gopher frogs breed most commonly during rainy periods from December into April, with the males generally arriving at the breeding sites and beginning to call before the females arrive. Heavy rains from tropical storms or hurricanes may stimulate late summer or fall breeding. Each female lays 500–3,000 or more eggs in shallow, fish-free seasonal wetlands, usually attaching them to live or dead vegetation, including small trees. Under natural conditions, the tadpoles may take from less than 3 to as many as 6 months to metamorphose into the juvenile frog

Eggs of a dusky gopher frog

A dusky gopher frog from southern Mississippi

stage, depending on water temperatures. Recently metamorphosed dusky gopher frogs are 1–1.5 inches long.

PREDATORS AND DEFENSE Suspected predators of adults include southern banded watersnakes, cottonmouths, predatory birds, and carnivorous mammals such as raccoons. Caddis fly larvae are known to eat the eggs and larvae, and other predaceous aquatic insects probably do as well. When handled, adult dusky gopher frogs inflate their body, lower their head, and put their front feet over their eyes. In addition, the warts on the skin secrete a pungent, whitish, bitter-tasting liquid.

CONSERVATION Dusky gopher frogs have succumbed to habitat degradation in the forms of urban development, agriculture, and forestry activities, and the species was listed as Endangered by the federal government in 2001. The only known remaining population at that time was at a single pond in the De Soto National Forest in Harrison County, Mississippi. In 2004 another population was located in Jackson County. The two populations are geographically isolated and have low genetic variability. Several conservation approaches have been proposed to optimize the habitat at the Harrison County site, including artificially mimicking natural fire regimes, translocating gopher tortoises to the terrestrial habitat around the wetland where the frogs breed, artificially increasing the length of time the pond holds water, moving eggs to other ponds, and creating new breeding ponds.

COMMENTS In 2006, amphibian biologists placed this species in the genus *Lithobates* and now refer to it as *Lithobates sevosus*.

TRUE TOADS

American toads are found throughout much of the eastern United States.

How do you identify an American toad?

SKIN
Rough

LEGS
Robust and stocky

FEET AND TOES
Hind toes without webbing

BODY PATTERN AND COLOR
Typically brown but may be reddish or gray; small dark spots

DISTINCTIVE CHARACTERS
Parotoid glands not flush against cranial crests; usually no more than two warts in spots on back

CALL
High-pitched trill lasting up to 30 seconds

SIZE
max tadpole = 1.5"
typical adult = 3"

American Toad *Bufo americanus*

DESCRIPTION American toads are basically dull brown, but individuals can range from reddish to olive, tan, or grayish. Dark splotches are sometimes present on the back; these may be yellowish or dark brown and seldom contain more than two small warts. A light line is sometimes present down the center of the back. The belly is white or cream colored and often has numerous dark spots or other dark markings. The average adult is 2–3.5 inches long; the largest females are up to 4.25 inches long.

WHAT DO THE TADPOLES LOOK LIKE? American toad tadpoles are small, dark brown or black, and reach a maximum length of slightly more than an inch. The tail fin is translucent, and the lower portion of the muscular part of the tail is lighter in color.

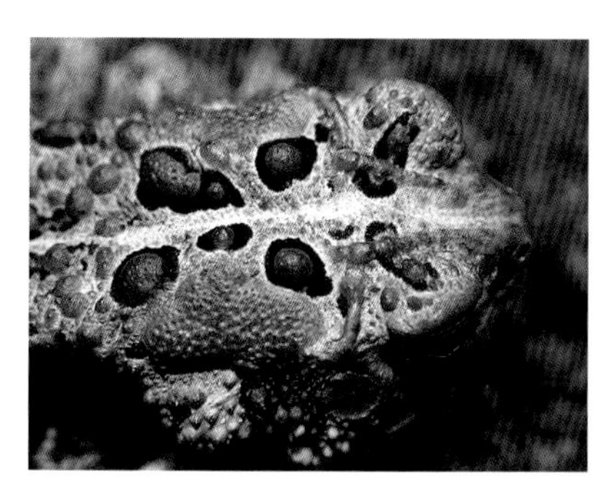

The American toad usually has fewer than three warts within each of the large dark spots on its back.

SIMILAR SPECIES American toads usually have only one or two warts within the large blotches on the back while Fowler's toads have three or more. Also, the cranial crests are not flush against the parotoid glands as they are on Fowler's toads, although a short projection from one or both cranial crests may touch the parotoid. American toads lack the cranial crest knobs found on southern toads, usually have more spotting on the belly than do Fowler's toads, and do not have light and dark stripes along the sides as Coastal Plain toads do. Oak toads are much smaller and have a more distinctive white stripe along the back.

DISTRIBUTION AND HABITAT American toads occur in some part of every southeastern state except Florida; they are found throughout most of Tennessee and Virginia and in noncoastal portions of the other states, and they are known or suspected from scattered localities in Louisiana. Two subspecies are recognized: the eastern American toad (*B. a. americanus*) and a western form, the dwarf American toad (*B. a. charlesmithi*). American toads interbreed with Fowler's toads and southern toads in some areas where their ranges overlap. This ubiquitous species occupies a variety of terrestrial habitats in the South-

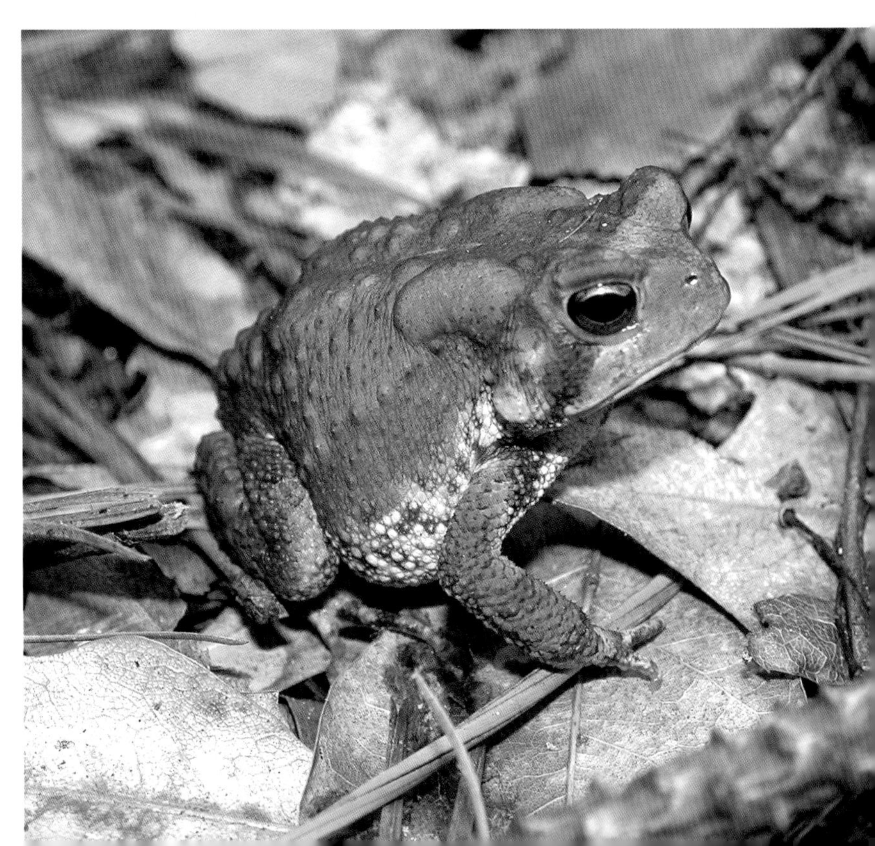

American toads are often gray or reddish in color.

east, including wooded areas, open fields and pastures, and suburban and agricultural areas.

BEHAVIOR AND ACTIVITY American toads are active primarily at night. They spend the day in hiding beneath logs, forest ground litter, rocks, or other cover objects, and often return to the same hiding spot each day. These toads are not territorial, but individuals remain within prescribed areas of several hundred square feet during the nonbreeding season. Adults will travel more than a half mile during and following rainy periods in order to breed. American toads remain dormant and under cover during hot, dry periods and during winter. They are seen infrequently during much of the summer and are generally completely inactive from October or November until breeding begins in the late winter or spring.

FOOD AND FEEDING Adults eat most kinds of small insects—especially ants, beetles, and moths—and other invertebrates. Individuals frequently sit beneath streetlamps or in other lighted areas and eat insects that come to the lights. The tadpoles eat a variety of aquatic organic matter, including algae, detritus, and dead fish or tadpoles.

DESCRIPTION OF CALL The advertisement call is a long, melodious trill that can last for as long as half a minute. Males also have a chirplike release call given when one male accidentally clasps another during mating. Females

Did you know?

Toads do not drink with their mouth; they absorb water through a patch of skin on their underside.

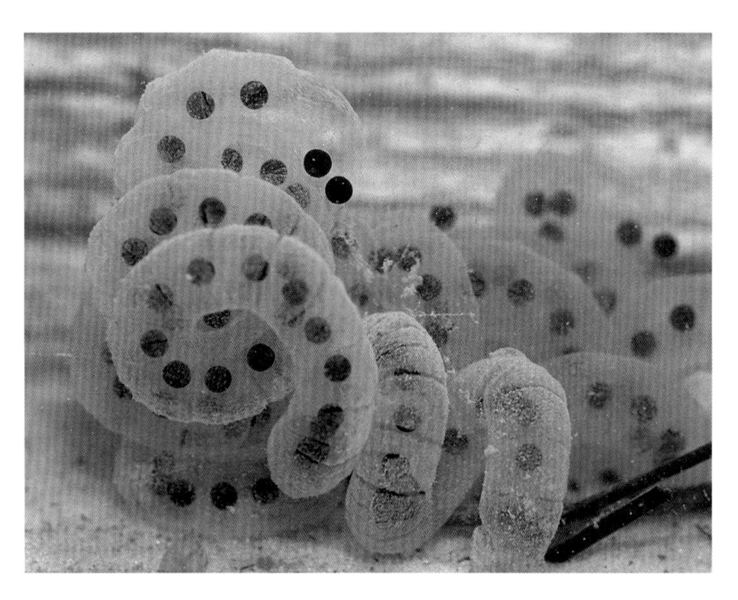

The eggs of American toads are laid in long, jellylike strings.

that have already mated and laid their eggs also produce this call if grasped by a male.

REPRODUCTION American toads begin breeding as early as January or February in the southern part of the geographic range and as late as March in the northern areas of the Southeast. During the breeding season males typically call at night, but they also sing on warm, wet days during the peak breeding period. Shallow, open aquatic sites, including ditches or borrow pits that are usually free of fish, are breeding sites. The enormous numbers of eggs, which are reported to range from 2,000 to 20,000 per female, are deposited in shallow water in a pair of gelatinous strings that are attached to vegetation or lie on the bottom at a depth of 2–4 inches. The eggs hatch in as few as 3 or as many as 12 days, and the tadpoles develop for up to 2 months before transforming into tiny toadlets 0.25–0.5 inch long. American toads are known to breed at 4 and 5 years of age and may live much longer.

PREDATORS AND DEFENSE Snakes (hognose, garter, and watersnakes), birds (ducks, crows, and screech owls), and mammals (raccoons, opossums, and striped skunks) are predators of adults. The tadpoles are eaten by predaceous diving beetles, giant water bugs, dragonfly naiads, crayfish, and birds (least sandpipers). American toads rely on their camouflage to avoid detection by predators and respond defensively by inflating the body. The toxic secretions of the parotoid glands and skin presumably protect adults from some predators. The eggs and tadpoles are toxic or distasteful to some aquatic predators. American toad tadpoles usually swim in dense schools.

CONSERVATION Because of their wide geographic range and their ability to thrive in most terrestrial habitats, including developed areas, American toads are not considered in need of special conservation measures in the Southeast.

COMMENTS In 2006, amphibian biologists placed this species in the genus *Anaxyrus*.

Some Fowler's toads appear reddish or greenish.

Fowler's Toad

Bufo fowleri

DESCRIPTION Fowler's toads are various shades of brown to gray, and some individuals appear greenish or reddish. Large, dark blotches that have dark edges and contain three or more warts are present on the body. The belly is usually white and unmarked. A white stripe typically extends down the center of the back. The lower front leg typically has no enlarged warts. The cranial crests form a ridge above each eye and touch the front end of the oval parotoid glands. Adults are typically 2–3 inches long; the maximum is usually less than 3.5 inches in the Southeast. When handled, these toads have a smell that some people describe as like raw peanuts.

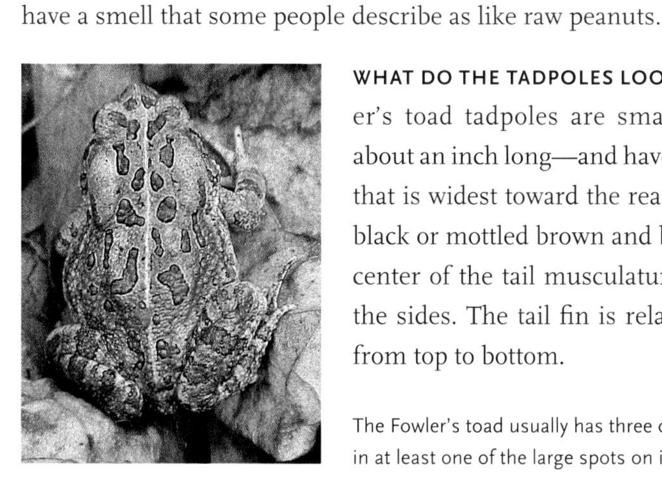

WHAT DO THE TADPOLES LOOK LIKE? Fowler's toad tadpoles are small—generally about an inch long—and have an oval body that is widest toward the rear. The body is black or mottled brown and black, and the center of the tail musculature is black on the sides. The tail fin is relatively narrow from top to bottom.

The Fowler's toad usually has three or more warts in at least one of the large spots on its back.

How do you identify a Fowler's toad?

SKIN
Rough

LEGS
Stocky

FEET AND TOES
Rear toes slightly webbed

BODY PATTERN AND COLOR
Brown or gray, sometimes reddish; large spots, belly white and unmarked

DISTINCTIVE CHARACTERS
Three or more warts in one or more large blotches on back; lower front legs without large warts

CALL
A *waaah* that lasts 1–4 seconds

SIZE
max tadpole = 1.5"
typical adult = 2.5"

Fowler's Toad

Bufo fowleri

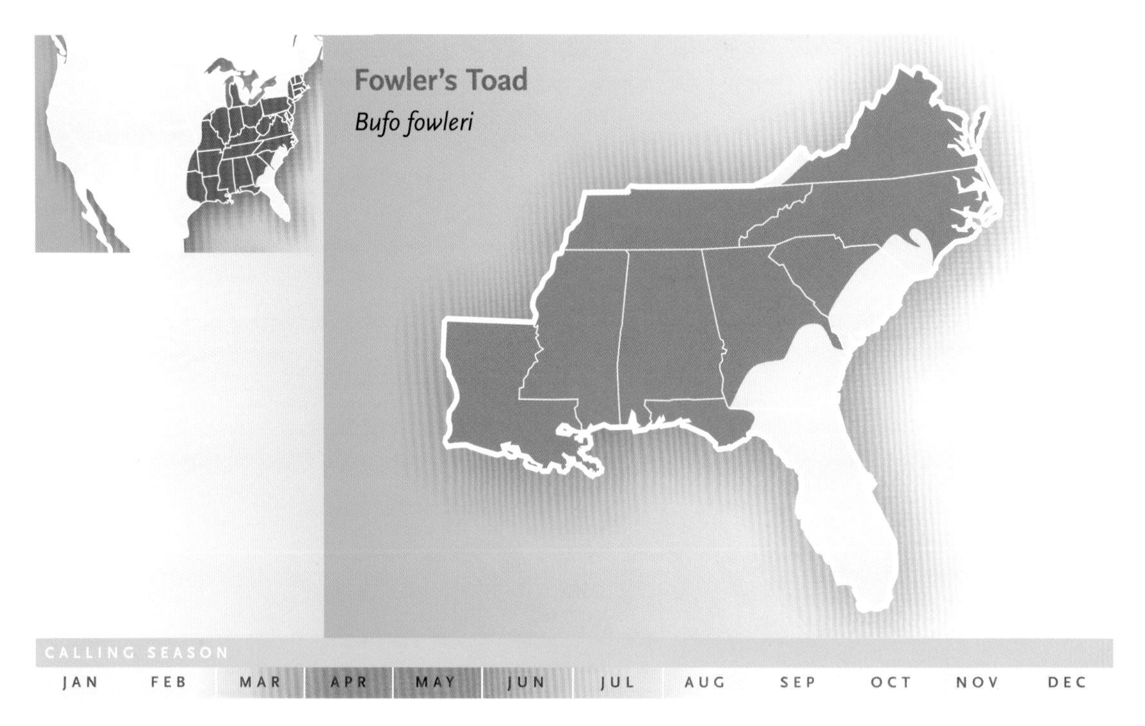

Fowler's toads occur in most habitats within their geographic range in the Southeast.

SIMILAR SPECIES Fowler's toads can be reliably distinguished from most similar-appearing true toads within their geographic range, although the identity of hybrid individuals can be difficult to determine. American toads typically do not have three or more warts within large, dark blotches on the back and a white belly with minimal spotting. Coastal Plain toads have a furrow between the cranial crests, and a sharply contrasting white line above a dark line and a distinctive row of enlarged tubercles on the sides. Southern toads usually have knoblike projections on the cranial crests. Oak toads are smaller, and the light-colored line down their back is more distinctive.

DISTRIBUTION AND HABITAT Fowler's toads occur in all or part of every southeastern state; they are absent from southeastern North Carolina, most of the Coastal Plain of South Carolina and Georgia, and the Florida peninsula. Fowler's toads hybridize with American toads, southern toads, and Coastal Plain toads in areas where their ranges overlap. Fowler's toads occur in most habitats within their geographic range in the Southeast, including coastal and low-elevation mountainous areas, woods and fields, and sandy or organic soils. They also can be common in urban or agricultural areas.

In areas where Fowler's toads and southern toads overlap, the former are reported to occupy bottomland habitats rather than the upland areas that are more characteristic of southern toads.

BEHAVIOR AND ACTIVITY Fowler's toads are active at night from spring through summer but also move about in the daytime during warm, rainy weather or otherwise moist conditions. Individuals will migrate more than 200 yards from their breeding sites to habitats where they spend most of the nonbreeding season. They are known to occupy home ranges of more than a half-acre and to return to the same area each year, but they do not attempt to exclude other toads. Fowler's toads avoid drought, extreme heat, and cold by burrowing into soft soil or sand, or by hiding in mammal burrows or tree root holes. During hibernation, their self-made burrows are often more than 6 inches below the surface.

FOOD AND FEEDING Fowler's toads are noted for eating many kinds of invertebrates, with a preference for ants and beetles and an apparent aversion to earthworms. The young toads eat flies, aphids, and springtails. Part of the adults' feeding strategy is to walk toward prey instead of hopping. The tadpoles generally eat small particles suspended in the water.

A male Fowler's toad calling to attract a mate. The dark throat distinguishes males from females.

DESCRIPTION OF CALL The 1- to 4-second *waaah* that is the advertisement call of Fowler's toad has been described as a "weird, wailing scream" or a nonmusical buzzing. Both sexes give a short release call that resembles a chirp when they are not receptive to mating or when picked up by a human.

REPRODUCTION Fowler's toads move from terrestrial habitats to aquatic sites and begin breeding as early as February–April in Florida; in Virginia they breed from April to July. The peak breeding period is in April or May in most parts of the Southeast, but breeding continues into the fall in some southern areas. Fowler's toads may breed later than American toads in localities where both species occur. Eggs are laid in shallow aquatic habitats that can vary from flooded ditches or ground depressions to shallow shorelines of lakes, ponds, and even rivers. Vegetation is usually present in the water. Females commonly lay about 4,000–5,000 eggs, but some clutches contain as many as 10,000. The eggs, which are enclosed in gelatinous strings that are either single or in pairs, hatch into tadpoles after about a week. Tadpoles metamorphose after 40–60 days into froglets that are less than half an inch long.

PREDATORS AND DEFENSE Eastern hognose snakes, loggerhead shrikes, American bitterns, bullfrogs, and raccoons are all known predators of adults. Toxic secretions from the parotoid glands and skin probably repel some mammal and bird predators. An additional defense strategy is camouflage against the soil substrate. The tadpoles are unpalatable to some fish, but eggs and tadpoles may fall prey to some aquatic salamanders and predatory insects.

CONSERVATION Fowler's toads are susceptible to certain pesticides and acid rain, but conservationists have not focused on the species because of its wide geographic range and abundance in many areas.

COMMENTS Some scientists consider toads in western Louisiana to be a separate species, Woodhouse's toad (*Bufo woodhousii*). However, other experts consider Fowler's toads to occur throughout the entire state of Louisiana, with Woodhouse's toads only occurring more westward. For purposes of this book, we refer to the species throughout Louisiana as Fowler's toad, with the understanding that its biological history and relationships with other true toads are complex. The East Texas toad is considered a subspecies (*B. w. velatus*) of Woodhouse's toad by some amphibian authorities, a separate species (*B. velatus*) by others, and merely a hybrid or variant of Woodhouse's toad by others. However, because Fowler's toad is known to hybridize in Louisiana with the Coastal Plain toad and the southern toad, the genetic relationship between Fowler's toad and the East Texas toad is also equivocal. In 2006, amphibian biologists placed this species in the genus *Anaxyrus*.

A Fowler's toad seeking refuge under pine needles

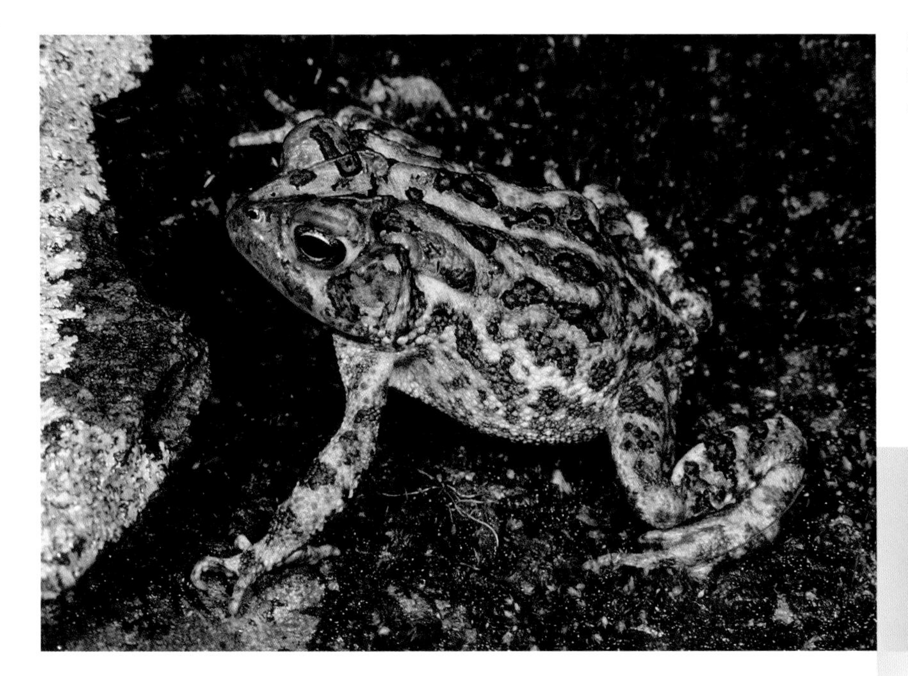

Southern toads are common throughout much of the Southeast.

Southern Toad

Bufo terrestris

DESCRIPTION The basic body color is brown, but it can range from dull reddish to almost black. Individuals in the same geographic area can vary greatly in color and appearance, and even the same individual may look different at different times. A light stripe extends partway down the center of the back of some individuals but is frequently absent. Several dark spots are often present on the back and sides, each surrounding one or two small warts. The cranial crests of adults are prominent and distinctive in having knobs that project upward in front of the parotoid glands. The belly is variable shades of grayish white. Typical body sizes of adults range from about 1.5 to 3.5 inches, but females from Ossabaw Island, Georgia, and Cat Island, Mississippi, and possibly other isolated island populations may reach lengths in excess of 4.5 inches.

WHAT DO THE TADPOLES LOOK LIKE? Southern toad tadpoles are dark brown or black above and below and reach a maximum length of about an inch. They are shaped like a teardrop, being broader to-

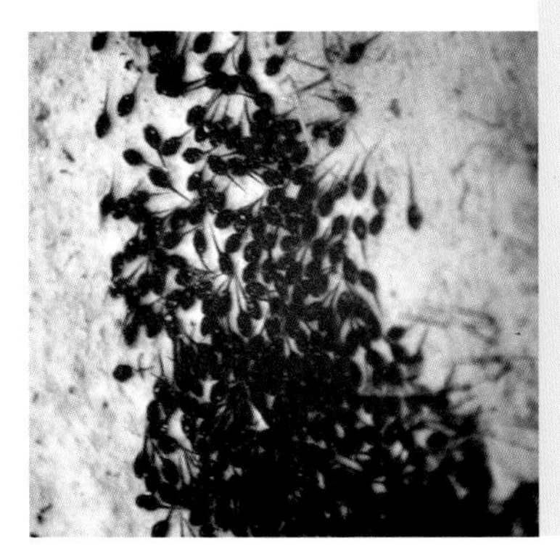

Southern toad tadpoles

How do you identify a southern toad?

SKIN
Rough

LEGS
Stout and short

FEET AND TOES
No webbing between hind toes

BODY PATTERN AND COLOR
Usually brown but may be reddish to black; belly cream colored

DISTINCTIVE CHARACTERS
Cranial crests have elevated knobs on top of head between eyes and parotoid glands

CALL
High-pitched trill lasting several seconds

SIZE
max tadpole = 1.5"
typical adult = 2.5"

Southern toads vary from brown to reddish brown.

ward the rear than many other tadpoles. An orangish mark behind each eye is oriented upward and backward. The tail fin is transparent with dark speckling that is heavier on the upper portion. Southern toad tadpoles sometimes form schools that swim in relatively open water.

SIMILAR SPECIES Southern toads could be confused with some of the six other southeastern true toads whose ranges overlap with theirs, but the prominent knobs on the cranial crests, usually no more than two warts within the dark blotches on the body, and high-pitched trill are distinctive. Southern toads hybridize with American toads and Fowler's toads in regions where their ranges overlap, and the offspring are intermediate in appearance, making clear identification difficult.

DISTRIBUTION AND HABITAT Southern toads are found in every southeastern state except Tennessee; they occur mostly below the Fall Line from southeastern Virginia south throughout Florida to eastern Louisiana, and in one isolated population in northwestern South Carolina. This species

Southern Toad
Bufo terrestris

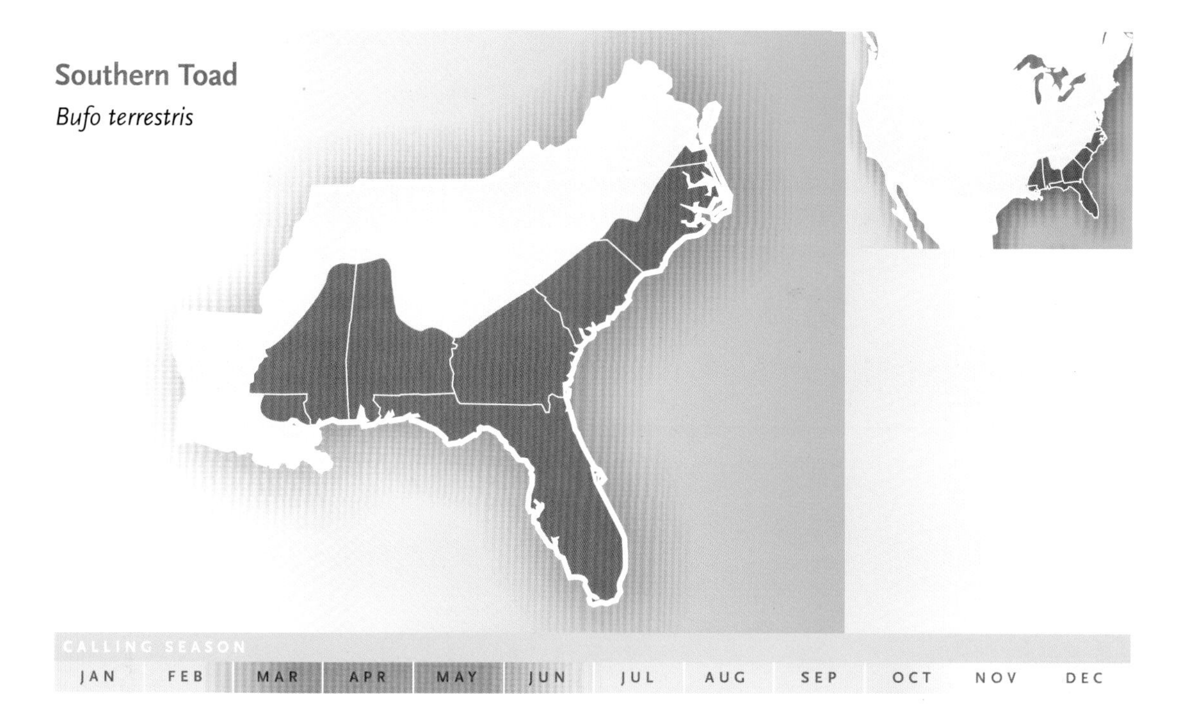

occupies a variety of terrestrial habitats that are typically associated with sandy soils, including pine forests, coastal scrub and forest habitats, and open fields. Southern toads usually persist under conditions of agriculture, forestry clear-cuts, and residential development as long as aquatic breeding sites are in the vicinity.

The cranial crests of southern toads characteristically form large knobs behind the eyes.

BEHAVIOR AND ACTIVITY Foraging southern toads are active primarily from dusk through the night during warm weather. They are active throughout the year in most of Florida and are inactive in late fall and winter in more northerly portions of their range. They travel extensively through the terrestrial habitat, and an individual can cover as much as a square mile moving between feeding, hibernation, and breeding sites. Toads often hide under forest litter or dig into loose soil during the day and bury themselves during winter cold spells or droughts.

FOOD AND FEEDING Southern toads will eat any animal they can catch and swallow. Among their known invertebrate prey are snails and a variety of insects (ants, beetles, earwigs, lightning bugs, mole crickets, honeybees, and roaches), but they will probably eat any insect that is available. The tadpoles primarily eat algae that they scrape from leaves and aquatic vegetation.

DESCRIPTION OF CALL The advertisement call is a high-pitched, melodious trill that lasts for several seconds. When captured, southern toads often emit a chirping release call while vibrating their body.

Southern toads may breed in any shallow water body.

REPRODUCTION Southern toads breed from as early as January to as late as October, with most of the reproductive activity occurring in early to late spring. During warm weather, adults move from upland terrestrial habitats to breeding sites in response to rains, and males may call on warm nights during most months of the year. Southern toads are somewhat indiscriminate in their choice of breeding sites but generally select areas of standing water such as water-filled ditches and tire ruts; flooded low areas in parking lots, fields, and woods; shallow swamp pools; Carolina bays and other isolated wetlands; and the edges of farm ponds. Each female lays 2,500–4,000 eggs, strewing them along the top of the water in long, gelatinous strands. The eggs hatch within 2–4 days, and the tadpoles undergo metamorphosis in 1–2 months when they are less than half an inch long. Southern toads can live for at least 10 years and presumably continue to breed each year.

PREDATORS AND DEFENSE Southern toads are eaten by several snake species, including eastern and southern hognose snakes, plain-bellied and southern banded watersnakes, black racers, garter snakes, and indigo snakes. Giant water bugs and turtles have been reported to kill and eat adults in aquatic areas during the breeding season. Predators on tadpoles include salamanders (amphiumas and sirens) and aquatic insects. Adults' primary defense on land is concealment beneath logs and ground litter or beneath sand or loose soil. When threatened by a predator, a southern toad will inflate its body and bend forward to expose the parotoid glands, which produce secretions that many animals find unpalatable or even poisonous. The eggs and tadpoles are also foul tasting to some aquatic predators.

CONSERVATION Southern toads are not protected in any part of their geographic range and seem to fare well in many suburban and agricultural areas. A noticeable decrease in southern toads has been reported in areas of Florida where cane toads have been introduced, although whether the two species compete for resources is undetermined. It is possible or even likely that cane toads actually prey on their smaller relatives.

COMMENTS In 2006, amphibian biologists placed this species in the genus *Anaxyrus*.

The Coastal Plain toad nearly always has a large dark stripe running along each of its sides.

Coastal Plain Toad *Bufo nebulifer*

DESCRIPTION Body color ranges from yellowish brown to dark gray, often within a single individual. Adults characteristically have a line of small but distinctive wartlike tubercles extending along each side next to a light stripe; these are bordered by a dark area above and by a larger dark stripe below. Adults typically reach 2–4 inches, with occasional females being more than 5 inches long. A cream-colored light stripe is often present down the center of the back. The belly is cream to grayish. The triangular parotoid glands often curve back and downward. The cranial crests are sharply defined, creating a furrow between them on the top of the head. Males have a dark greenish throat.

WHAT DO THE TADPOLES LOOK LIKE? Coastal Plain toad tadpoles are small and black and may have gold flecking. Several black bars on the top half of the tail musculature alternate with lighter bars. The tail fin is transparent with dark mottling on the upper portion and possibly elsewhere. Tadpoles transform into toadlets less than half an inch long.

SIMILAR SPECIES Coastal Plain toads are most readily distinguished from all other southeastern toads by the distinctive light stripe extending down the side with the broad dark stripe beneath it. The furrow between the cranial crests is also characteristic of the species.

How do you identify a Coastal Plain toad?

SKIN
Rough

LEGS
Short and stocky

FEET AND TOES
Hind toes slightly webbed

BODY PATTERN AND COLOR
Yellowish brown to dark gray, usually a light stripe down the back; dark-bordered, light line extends down the sides

DISTINCTIVE CHARACTERS
Light longitudinal side stripe bordered by broader dark stripe below; prominent cranial crests forming furrow between eye

CALL
Short, harsh trills

SIZE
max tadpole = 1.5"
typical adult = 3.5"

DISTRIBUTION AND HABITAT Coastal Plain toads are found across the southern half of Louisiana, throughout southeastern Texas, and in scattered counties in southern Mississippi; isolated populations are present in northeastern Louisiana. The Coastal Plain toad interbreeds with Fowler's toad in some areas where the two species overlap, and although survivorship of the offspring may be high, many are sterile. These toads occur in most lowland and moist terrestrial habitats within their range and appear to thrive in residential areas as well. The species is less prevalent in pinewoods than some of the other true toads, and is more likely to be found in agricultural, urban, or hardwood habitats. They are seldom limited by breeding sites in the Southeast as they typically live in areas with abundant rainfall and are capable of breeding in roadside ditches, borrow pits, and other man-made or natural standing water.

Coastal Plain toads have a large furrow on top of the head.

BEHAVIOR AND ACTIVITY Coastal Plain toads are active mostly at night, when they leave their daytime hiding places to forage. Some individuals climb a short way up oak trees and hide in tree holes to which they return after feeding. Adults have been found living within 100 feet of saltwater habitats. Coastal Plain toads are capable of moving rapidly within terrestrial habitat, and can travel up to 0.75 mile in 2 days. Most probably stay within an area with a radius of about 150 feet, although they do not defend territories. These toads are inactive during droughts and cold weather, hiding beneath natural ground litter and artificial cover or in rodent burrows.

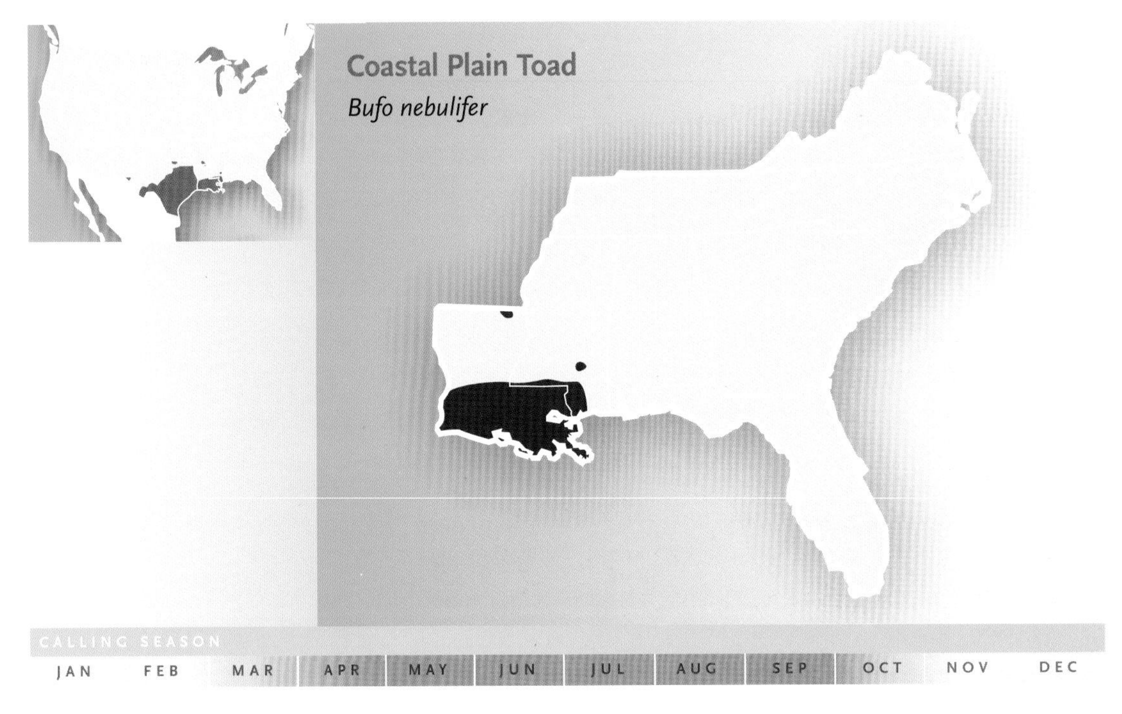

FOOD AND FEEDING Coastal Plain toads eat insects and other invertebrates, including beetles, isopods, and scorpions. A fence lizard and young Fowler's toad have also been reported as prey, suggesting that adults will eat anything small enough to subdue and swallow. The tadpoles eat algae that they scrape off submerged plants or other underwater surfaces.

DESCRIPTION OF CALL The advertisement call is a series of short trills that are lower pitched and do not have the melodious quality of the southern toad's call. Males give a short release call best described as fast chirping when grasped accidentally by another male during the courtship and breeding period.

REPRODUCTION Coastal Plain toads in the Southeast seem to find any still, shallow water suitable for breeding, and they commonly breed in flooded ditches and other artificial bodies of standing water as well as in river floodplains. Breeding in slightly brackish water has also been reported. Adults come to breeding sites after heavy rains, typically from March through August. A female may lay as many as 20,000 eggs, typically in shallow water without vegetation. The eggs are in gelatinous strings, often in paired rows. They hatch within 2 days, and the tadpoles develop rapidly, metamorphosing within 3–4 weeks into froglets that are about half an inch long. Individuals have been known to live 8 years in the wild.

Coastal Plain toads can be common in suburban areas and agricultural areas as well as in hardwood forest habitats.

PREDATORS AND DEFENSE Western ribbon snakes, diamondback watersnakes, and indigo snakes are all known predators, and eastern hognose snakes are likely predators. Tiger salamanders probably eat young toads, and diving beetles have been reported to eat tadpoles. The poisonous secretions of the parotoid glands of adults discourage some predators, and the eggs are toxic as well.

CONSERVATION Because of their ability to persist in a variety of habitats, including agricultural and urban areas, the Coastal Plain toad is not considered a species of conservation concern.

COMMENTS The Coastal Plain toad, *Bufo nebulifer*, was known as the Gulf Coast toad, *Bufo valliceps*, until DNA studies in the late 1990s revealed that the populations in the Southeast and in northeastern Mexico were distinct from populations in southern Mexico that extend into Central America. The name *B. valliceps* was applied to the Mexican and Central American species. In 2006, amphibian biologists placed *B. nebulifer* in the genus *Anaxyrus*.

The oak toad characteristically has a well-defined light stripe down the middle of the back.

How do you identify an oak toad?

SKIN
Rough

LEGS
Relatively short

FEET AND TOES
Hind toes unwebbed

BODY PATTERN AND COLOR
Gray with brownish markings on either side of a whitish or yellowish stripe; belly whitish

DISTINCTIVE CHARACTERS
Small size; light-colored line down back

CALL
Repeated, high-pitched peeps sounding like a baby chicken

SIZE
max tadpole = 1.5"
typical adult = 1"

Oak Toad

Bufo quercicus

DESCRIPTION Oak toads are the smallest true toads in the United States; the largest individuals are less than 2 inches long. The body color is usually light to dark gray, with an occasional brownish tint. Large, widely spaced brown or black blotches on the back sometimes form broad lines along the sides. A very distinct light line that is often white but may be yellowish extends down the center of the back. The gray legs have broad, dark bands. The cream-colored belly has a granular appearance. The parotoid glands are large and extend from either side of the back of the head down the neck, but they may be less obvious than those of other toads because the oak toad is so small.

The adult oak toad is only slightly more than 1 inch long.

WHAT DO THE TADPOLES LOOK LIKE? Like the adults, the tadpoles are small, usually less than 1 inch long, and may be gray, dark greenish brown, or almost black with a whitish belly. The mouth and snout turn downward. The tail fin is transparent, but the central, muscular area of the tail appears to have light and dark bands.

Oak Toad

Bufo quercicus

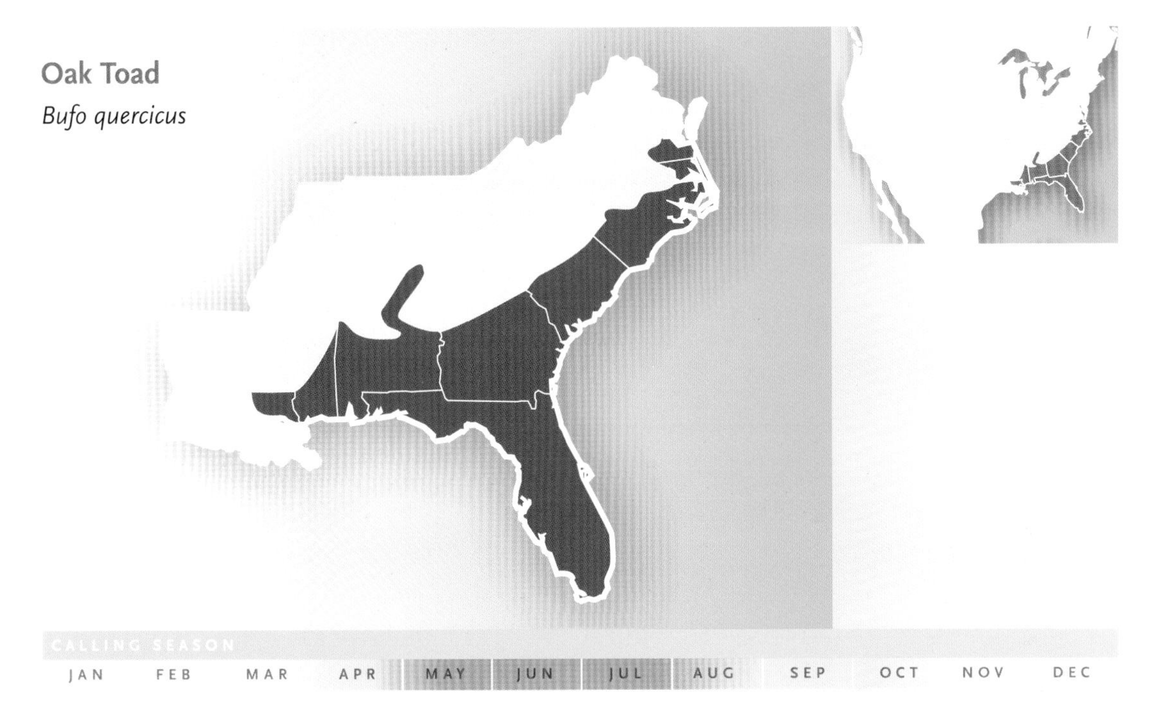

SIMILAR SPECIES Oak toads are readily distinguished from most other toads by their small size and the prominent light center stripe, which contrasts sharply with the gray body. Southern toads occasionally have a hint of a light line down the back, but they have cranial crests in front of their parotoid glands. Coastal Plain toads with a light stripe down the center also have a distinct light-colored stripe down each side with a broad, dark stripe below it.

DISTRIBUTION AND HABITAT Oak toads are found only in the Southeast, ranging throughout Florida and in parts of every other southeastern state except Tennessee. Their distribution is limited primarily to the Coastal Plain but extends north into central Alabama. Oak toads are typically found in association with pine woods or mixed oak and pine habitats, often with grassy areas. They breed in a variety of wetland situations including Carolina bays, freshwater marshes, cypress ponds, and flooded ditches and low areas.

BEHAVIOR AND ACTIVITY Like most other frogs and toads, oak toads are active at night, but they are also active—and most often observed—during the day, especially under moist conditions during the warmest months. During unfavorably dry or cold weather they retreat to underground burrows or beneath logs and other ground litter. They sometimes live in terrestrial habitats as much as 250 feet from the wetland breeding sites to which they migrate during rainy periods. They are inactive during cold weather from late fall to early spring.

Did you know?

Toads have warts, but handling a toad cannot give a person warts.

An oak toad calling

FOOD AND FEEDING Oak toads eat a wide variety of small insects and other invertebrates, including beetles, flies, spiders, and centipedes, but are particularly noted for consuming ants. The tadpoles eat algae and decaying organic material.

DESCRIPTION OF CALL The advertisement call sounds like the peeping of a baby chick and is repeated continuously with short pauses. The call is very loud for such a small toad, and large choruses can often be heard 200 or more yards away.

REPRODUCTION Oak toads breed predominantly in the spring, but if temperatures stay warm and wet they will continue to call and lay eggs until early autumn. Breeding takes place in shallow water, especially in flooded depressions in pine flatwoods, but also in water-filled ditches or pools in cypress swamps. The female lays 300–500 eggs in short strings of only a few eggs each. Eggs are usually laid 2–5 inches below the water's surface and are attached to grass or other vegetation. They hatch in a day or a day and a half, and the tadpoles typically metamorphose after about 2 months. The tadpoles are tiny and seldom reach an inch in length. Young oak toads are also tiny, only about a quarter of an inch long at metamorphosis.

PREDATORS AND DEFENSE Raccoons, snakes (hognose and garter snakes), other frogs (gopher frogs), and toads (cane toads) are known predators. The parotoid glands produce secretions that are probably toxic to some would-be predators, and the eggs are also reported to be toxic.

CONSERVATION Because they occur in localized populations, the most obvious threat to the well-being of oak toads is habitat destruction through the suppression of natural fires; the destruction of pine woods habitats; and the loss of small, fish-free, ephemeral wetlands as breeding sites.

COMMENTS In 2006, amphibian biologists placed this species in the genus *Anaxyrus*.

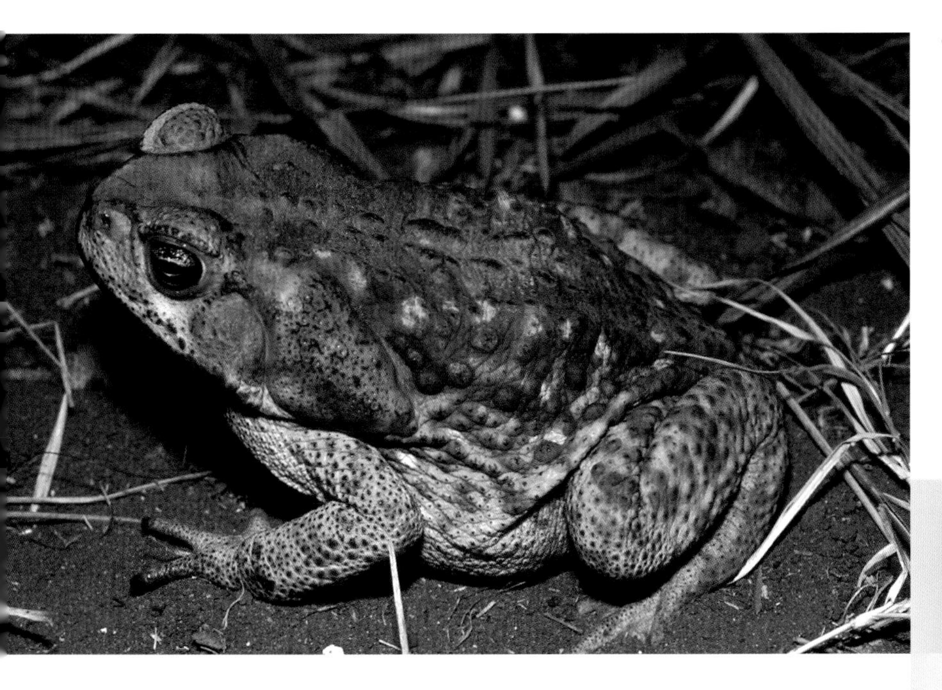

Cane toads were first introduced into south Florida in the 1930s.

How do you identify a cane toad?

SKIN
Rough and warty

LEGS
Short and robust

FEET AND TOES
Rear toes slightly webbed

BODY PATTERN AND COLOR
Back brown to gray; belly yellowish white

DISTINCTIVE CHARACTERS
Large size; prominent parotoid glands that extend down the sides

CALL
Continuous low-pitched, slow trill

SIZE
max tadpole = 1.5"
typical adult male = 4"
typical adult female = 5"

Cane Toad

Bufo marinus

DESCRIPTION Cane toads are the largest toads in the Southeast; adult body lengths commonly range from 3 to 6 inches, and some females in Florida reach lengths approaching 7 inches and can weigh more than 3 pounds! The basic dull brown body color can vary from grayish to reddish. Some individuals have dark markings. Warts are prominent all over the back and sides. The light-colored belly appears grainy and may have a yellowish tinge. The parotoid glands are large, extending behind the head downward to the level of the jaw line. Cranial crests extend from behind each eye, and the lateral ridges contact each parotoid gland. The back of breeding males takes on a cinnamon hue and a sandpaper-like texture. Adult females are larger than adult males.

WHAT DO THE TADPOLES LOOK LIKE? The small (less than 1.5 inches) black tadpoles give no indication of the ultimate size they will reach as adults. The black tail musculature is visible down the center of the transparent tail fin.

SIMILAR SPECIES Their large size easily distinguishes adult cane toads from our native toads. Smaller individuals might be confused with other toads, but the large parotoid glands extending from behind the eyes and down the sides are diagnostic.

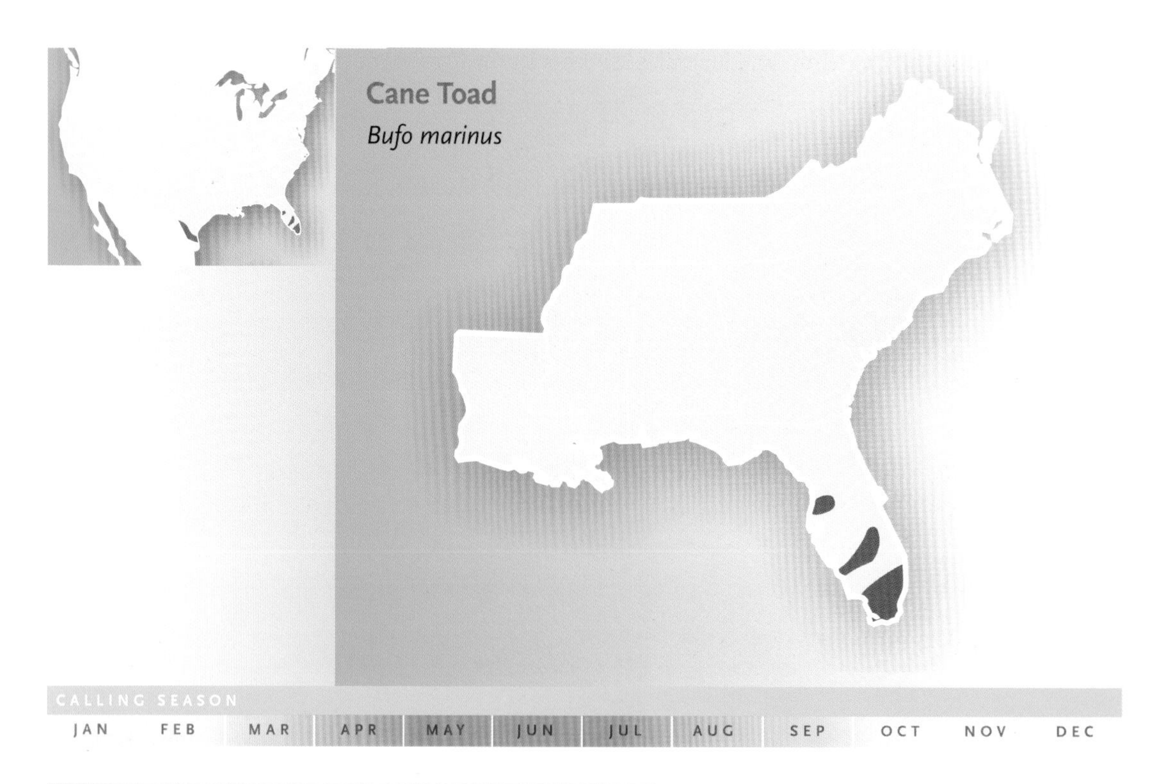

Cane toads have very large parotoid glands that secrete a powerful toxin when the toad is attacked.

DISTRIBUTION AND HABITAT Cane toads occur naturally only from southern Texas to South America. They are an invasive species in many parts of the world, though, most notably in Australia and Japan, as well as the Miami, Florida, area, where they became established sometime before 1955. They were intentionally released in southern Florida in the 1930s and 1940s to control pests in sugarcane fields, but these early populations are believed to have died out. By the 1960s cane toads were prevalent in much of the southern tip of Florida, and in the following decades they became established in several counties midway up the Florida peninsula. By 2004 the species had been recorded as far north as Clay County, Florida, southwest of Jacksonville. Cane toads can persist in most habitats that have sufficient moisture, but they appear to do especially well in agricultural fields and urban housing developments. Their absence from fully natural habitats has been noted in southern Florida. Cane toads are not aquatic, but they usually remain within a few hundred feet of a permanent water body.

BEHAVIOR AND ACTIVITY Adults and juveniles are primarily active at night; young toadlets move about during the day. These toads are mostly terrestrial but will occasionally climb into small trees. They are active throughout the year in southern Florida but primarily from March to November. They are inactive during cold spells and are reported to die when exposed to temperatures below 40° F for several days. They can tolerate some desiccation, but during extended dry periods they seek shelter under leaf litter, logs, rocks, or the soil itself. Cane toad tadpoles prefer warm water but can tolerate temperatures as low as 46° F. They travel in large schools that may include thousands of individuals.

Two cane toads in amplexus preparing to lay and fertilize eggs

FOOD AND FEEDING Cane toads prey on a wider variety of animals than any other toad or frog native to the Southeast. The list of invertebrates they will eat is extensive and includes beetles, ants, bees, roaches, spiders, scorpions, centipedes, millipedes, crabs, snails, and slugs. They also have been documented to eat squirrel treefrogs, oak toads, southern toads, ribbon snakes, ringneck snakes, and small mammals. On warm nights cane toads will move from their daytime shelters to outdoor lights, where they feast on the insects the lights attract. In Florida, they have been reported to eat dog food left on patios and porches. The tadpoles eat algae predominantly but will also eat smaller cane toad tadpoles.

Did you know?

*The world's largest frog, the West African goliath frog (*Conraua goliath*), can reach 7 pounds, about twice the size of an American bullfrog.*

DESCRIPTION OF CALL The advertisement call is a slow, steady, low-pitched trill of about 12 notes per second. The sound has been described as purring, drumming, or like someone tapping rapidly on wood.

REPRODUCTION Cane toads in Florida migrate from nearby terrestrial habitats to ditches, small puddles, and man-made canals after rains. They can breed in slightly brackish waters. Males are not territorial but will call throughout the year. Eggs are generally laid from March until September. A female can lay an amazing number of eggs—at least several thousand and sometimes as many as 20,000–25,000. The eggs are typically in long, gelatinous strings attached to vegetation in the water. The tadpoles develop rapidly, hatching from the egg in 2–3 days and transforming from the tadpole stage 2–4 weeks later. The young toads are less than half an inch long at metamorphosis.

Cane toads call from a variety of water bodies.

PREDATORS AND DEFENSE Cane toads are extremely toxic to many native predators and are toxic to some degree at every life stage from egg to adult. Dogs and cats can actually die from biting into the poisonous parotoid glands. Nonetheless, juveniles and small adults have numerous predators: birds such as crows, red-shouldered hawks, blue jays, and mockingbirds; and snakes such as hognose, garter, indigo, and southern banded watersnakes. Larger cane toads will eat smaller ones. The eggs and tadpoles are highly toxic to some fish and snails but not to some aquatic insects such as dragonfly larvae or to some crayfish. Yellow bullhead catfish native to Florida have been known to eat the tadpoles. Among the primary predators on tadpoles are larger tadpoles of the same species.

CONSERVATION The cane toad is a target of conservation measures not for its own protection but for the well-being of native species, both potential prey and would-be predators that can be poisoned by its toxic skin secretions. Because cane toads have been introduced into Florida, many native species are probably unaware of the potentially lethal toxicity of the secretions of the parotoid glands. Cane toads are likely to affect native toads and frogs directly by preying on them and perhaps indirectly by competing for the same food. The value of the species as a pest control agent, the original reason for bringing the cane toad to southern Florida, has not been confirmed. Further introduction of cane toads into southeastern habitats for pest control or as releases from the pet trade should be discouraged.

COMMENTS The cane toad is also called marine toad or giant toad in some parts of the world. In 2006, amphibian biologists placed this species in the genus *Chaunus*.

OTHER FROGS AND TOADS

A spadefoot toad from near Aiken, South Carolina

Eastern Spadefoot Toad *Scaphiopus holbrookii*

DESCRIPTION Eastern spadefoot toads can be brown, olive, gray, or black, but all have yellow markings shaped like reverse parentheses that extend the length of the body. Additional yellow markings may be present on the sides. The body is smooth and moist with several very small bumps, warts, or tubercles, many with an orange tinge. The belly is white or gray, and sometimes reddish toward the rear. A pair of glands is visible on the chest. The parotoid glands are not as prominent as those of the true toads. The pupils are elliptical like those of a cat. A sharp, dark, elongate spade used for digging is present on the inside of each hind foot. Adults are typically 1.5–2.5 inches long; some females are almost 3 inches long.

WHAT DO THE TADPOLES LOOK LIKE? The tadpoles seldom exceed 1.5 inches in length. The basic brown body color ranges from dark to bronze, sometimes with tiny orange spots. The tail fin is translucent, and the tail musculature has no dark markings. The internal organs, including gills, are visible through the translucent belly. The eyes are much closer together than the eyes of other southeastern tadpoles. The toadlets are slightly over half an inch

A spadefoot toad tadpole

How do you identify an eastern spadefoot toad?

SKIN
Smooth with small tubercles

LEGS
Short and stocky

FEET AND TOES
Distinct dark, spadelike protuberance on each hind foot

BODY PATTERN AND COLOR
Brown or dark gray; two irregular, inward-curving yellow lines on back

DISTINCTIVE CHARACTERS
Spadelike protuberance on each hind foot; vertical pupils; musky smell when handled

CALL
Loud *qwaa* repeated every 3–4 seconds

SIZE
max tadpole = 1.5"
typical adult = 2"

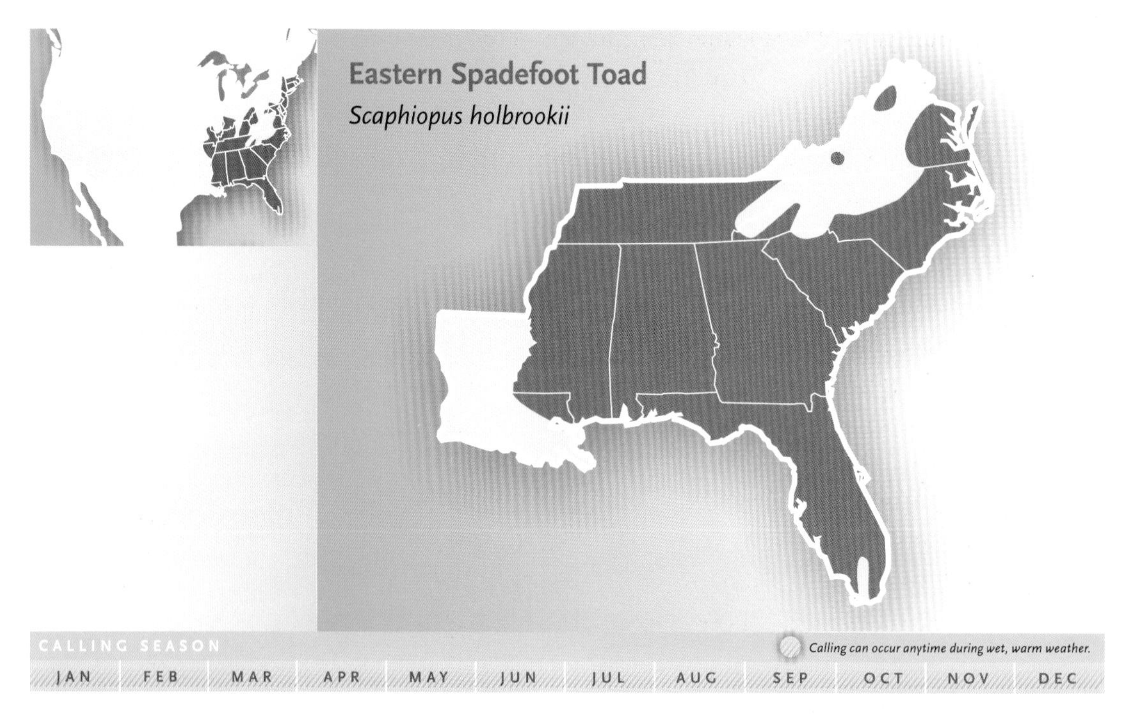

Vertical pupils (left) help distinguish spadefoots from other toads.

The "spade" on the hind foot (right) is used for digging and gives the spadefoot toad its name.

long. Spadefoot toad tadpoles do not hide under cover on the bottom as many tadpoles do but swim around openly, sometimes in schools of more than 10,000 tadpoles.

SIMILAR SPECIES The toadlike body shape differentiates spadefoot toads from true frogs and treefrogs. True toads and narrow-mouthed toads do not have the elliptical pupils, musky smell, or spadelike protuberances on the hind feet. Hurter's spadefoot toads have a raised area on the head between the eyes, and their geographic range does not overlap that of the eastern spadefoot.

DISTRIBUTION AND HABITAT Eastern spadefoot toads are found in all the southeastern states, south to the Florida Keys and west to eastern Louisiana, but are absent from parts of the Piedmont and higher-elevation mountains of Virginia and North Carolina. They occupy a variety of upland pine and hardwood forests as well as bottomlands, agricultural fields and pastures, and even open parks in urban areas. A common feature of all of the habitats is the presence of soft loam or sandy soils suitable for burrowing.

BEHAVIOR AND ACTIVITY Eastern spadefoot toads in the Southeast can be active in any month when temperatures are warm and humidity is high. Individuals characteristically spend most of the year in underground burrows that they create themselves by digging backward with the "spades" on the hind feet. On warm nights, even during winter in southern Florida, spadefoot toads emerge from their burrows after dark to feed. A spadefoot toad can go many weeks without feeding, however, if conditions are not suitable. They are cautious about leaving their burrows, to which they retreat rapidly if threatened. A burrow may be only a temporary or short-term retreat for some individuals, but others may use the same burrow for more than 4 years. Eastern spadefoot toads are not known to be territorial, but they generally do not share their burrow with other individuals. Home range size averages about 100 square feet, but these toads spend most of their time either inside their burrow or foraging in its immediate vicinity. Recently metamorphosed toadlets are noted for emerging in enormous numbers around the edge of breeding sites and can be seen commonly in the daytime. Even at that age they will travel more than 600 feet into the terrestrial habitat. Juveniles in the Southeast are believed to begin creating their own burrows within a month to 6 weeks after they transform from tadpoles.

When threatened, spadefoot toads may curl into a ball.

FOOD AND FEEDING Eastern spadefoot toads feed at night on a wide variety of nocturnal invertebrates including earthworms, millipedes, snails, moths, spiders, and beetles. Tadpoles eat both plant and animal plankton in the water; graze algae attached to underwater surfaces; scavenge on dead animals in the water, including other tadpoles and invertebrates; and will even eat the eggs of their own species.

DESCRIPTION OF CALL The advertisement call is a deep, short croak that has been variously described as *wank, wonk, quank, quonk, qwaah, owwww,* and *waaank* and is repeated continually at 3- to 4-second intervals. When spadefoot toads call in a large chorus, the sound is constant and can be

Spadefoots can breed year-round, but only during or after heavy rains.

heard more than a mile away. When grasped in a breeding aggregation, males will emit a quick series of raspy, low-pitched grunts.

REPRODUCTION Breeding is activated by heavy rains, and the species is known to breed explosively during any month in the Southeast as long as temperatures are warm enough, usually above 50° F. Males call en masse at night, and many will continue to call during the day if environmental conditions remain optimal. Individuals will travel several hundred yards from their underground burrows to aquatic breeding sites and then return to the same burrow. Eastern spadefoot toads can breed successfully in ditches, borrow pits, and shallow depressions in cleared fields that are temporarily flooded by rains; in artificial ponds (including suburban koi ponds); and in natural temporary ponds that fill during rains. Populations may go for years without breeding if heavy rains do not occur during warm weather. When suitable conditions do occur, a few females can produce thousands of successful offspring. Each female lays 2,000–5,000 eggs, attaching them to live or dead vegetation in the water. Eggs can develop within a single day during warm weather or may take as long as 2 weeks in the winter, and the tadpole stage lasts 2–8 weeks. Eastern spadefoot toads can live for a decade or more and continue to reproduce throughout their lives.

PREDATORS AND DEFENSE Spadefoot toads have a variety of predators, including amphibians such as southern toads and bullfrogs; reptiles such as cottonmouths, eastern and southern hognose snakes, banded and northern watersnakes, and black racers; birds such as gulls and cattle egrets; and mammals such as opossums and raccoons. Starlings and grackles will capture and eat young spadefoot toads around the margins of wetlands. Snapping turtles prey on the tadpoles, and fish and aquatic invertebrates are probably tadpole predators in some situations. Primary defenses of adults include inflating the body and curling forward into a tight ball. They also have a distinctive musky odor when captured, indicating the presence of skin secretions that may be toxic or distasteful. If transferred from a person's hand to eyes or mucous membranes of the nose, these secretions cause a burning sensation that may last for more than an hour.

CONSERVATION Eastern spadefoot toads are considered Rare, of Special Concern, or Endangered in parts of their geographic range outside the Southeast. In part because of the sporadic appearances of this species and the difficulty of determining distribution patterns and abundance on a local scale in most parts of the Southeast, clear conservation goals have not been established. Paving and building construction in areas where local parks or other open areas provide suitable terrestrial and breeding habitats undoubtedly reduce the size and number of local populations. Equally important is the fact that spadefoot toads cannot burrow through densely packed grass sod such as that found on lawns, further reducing the amount of habitat available for the species in residential areas.

Spadefoot toads can remain below ground for months. This sandy individual likely recently emerged from underground.

COMMENTS The common name for this group has traditionally been "spadefoot toads," although some amphibian biologists prefer the shortened name "spadefoot." See "Comments" in the Hurter's spadefoot toad species account on the relationship between the two species of spadefoot toads native to the Southeast.

A spadefoot toad from South Carolina

A Hurter's spadefoot toad from western Louisiana

How do you identify a Hurter's spadefoot toad?

SKIN
Smooth with small tubercles

LEGS
Short and stocky

FEET AND TOES
Dark, spadelike protuberance on each hind foot

BODY PATTERN AND COLOR
Dark green to dark brownish with irregular yellow or white lines down the back

DISTINCTIVE CHARACTERS
Spade on each hind foot; bump between eyes; vertical pupils; musky smell when handled

CALL
Short, guttural *qwaa* repeated every 1–2.5 seconds

SIZE
max tadpole = 1.5"
typical adult = 2"

Hurter's Spadefoot Toad *Scaphiopus hurterii*

DESCRIPTION Hurter's spadefoot toad looks very much like the eastern spadefoot with minor differences such as the basic color being dark green grading from brownish to black and the yellow lines extending down the back sometimes being more white. The smooth, moist body has numerous small tubercles, and the belly is white or gray with a conspicuous pair of glands in the chest area. The parotoid glands are elongate and sickle shaped. Most individuals have a distinctive raised area on the head between the eyes. The pupil of the eye is vertical. The inside of each hind foot bears a dark spade used for digging. Adults are typically 1.75–3 inches long, with some females reaching a length of 3.25 inches.

WHAT DO THE TADPOLES LOOK LIKE? The small, brown tadpoles sometimes have small orange spots. The tail fin is translucent, and the tail musculature lacks dark markings. The internal organs, including gills, are visible through the translucent belly. Like eastern spadefoots, Hurter's spadefoot tadpoles have eyes that are much closer together than those of other southeastern tadpoles. Tadpoles metamorphose as toadlets at a body length slightly over one inch. Also like other spadefoots, Hurter's spadefoot tadpoles swim around openly, sometimes in schools numbering more than 10,000.

Hurter's Spadefoot Toad
Scaphiopus hurterii

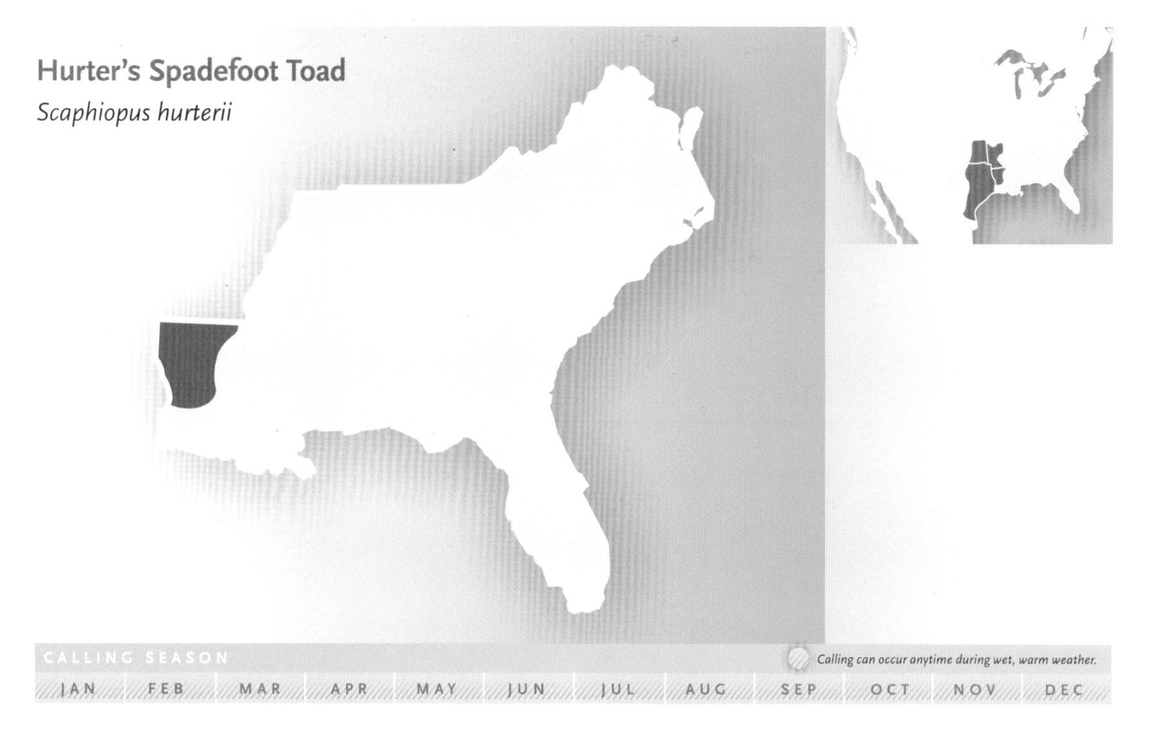

SIMILAR SPECIES Eastern spadefoot toads do not have the raised area on the head between the eyes that is characteristic of Hurter's spadefoot toads. Also, the geographic ranges of the two species do not overlap. The toadlike body shape differentiates Hurter's spadefoot toads from true frogs and treefrogs. Fowler's toads and narrow-mouthed toads do not have elliptical pupils, a musky smell, or spades on the hind feet.

Hurter's spadefoot toads closely resemble eastern spadefoot toads but have a raised area on the head between the eyes.

DISTRIBUTION AND HABITAT The geographic range of Hurter's spadefoot toad in the Southeast is restricted to the north-central portion of Louisiana. The species is also found in portions of Arkansas and in eastern Oklahoma and Texas. In Louisiana, this species is associated with woodland or open habitats with loose soil suitable for burrowing.

BEHAVIOR AND ACTIVITY Hurter's spadefoot toads can be active in any month when temperatures are warm. They spend most of the year in burrows they dig themselves with the spadelike protuberances on the hind feet. They rarely venture far from their burrows and do so only at night. Recently metamorphosed Hurter's spadefoot toads often emerge in large numbers from breeding sites and may be seen in the daytime as they disperse hundreds of yards into the terrestrial habitat. Juveniles probably begin creating their own burrows within a month to 6 weeks after they transform from tadpoles.

FOOD AND FEEDING Hurter's spadefoot toads feed at night, presumably on nocturnal invertebrates such as earthworms, millipedes, snails, moths, spiders, and beetles.

DESCRIPTION OF CALL The advertisement call is a guttural note that lasts a half-second and is repeated every 1–2.5 seconds. Some herpetologists describe the call as having a milder, less harsh quality than the call of the eastern spadefoot toad; others consider the two calls almost identical.

REPRODUCTION Hurter's spadefoot toads breed explosively during and following heavy rains. Their breeding characteristics are presumed to be similar to those of the eastern spadefoot toad (see species account), although detailed studies are lacking. Hurter's and eastern spadefoot toads can interbreed successfully when placed together.

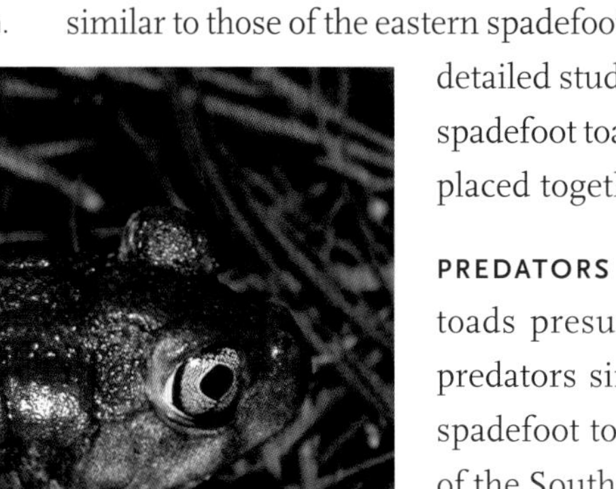

The Hurter's spadefoot toad is dark green to dark brownish.

PREDATORS AND DEFENSE Hurter's spadefoot toads presumably have an array of vertebrate predators similar to those that prey on eastern spadefoot toads in Louisiana and in other parts of the Southeast. These include larger toads and bullfrogs, eastern hognose snakes, banded and northern watersnakes, cottonmouths, gulls, egrets, opossums, and raccoons. Their defenses include inflating the body, curling into a tight ball, and secreting skin toxins that have an unpleasant smell or taste.

CONSERVATION Hurter's spadefoot is susceptible to pesticides, but there is no reason to suspect that it is any more vulnerable to them than other species of frogs and toads are. No special conservation measures have been taken for the species in its southeastern range in Louisiana.

COMMENTS Although Hurter's spadefoot was described as a distinct species in 1910, most herpetologists considered it a subspecies of the eastern spadefoot during much of the latter half of the 1900s. The two were again recognized as separate species in the early 2000s.

The pear-shaped body of the eastern narrowmouth toad ends in a noticeably pointed nose.

Eastern Narrowmouth Toad *Gastrophryne carolinensis*

DESCRIPTION The pear-shaped body of this toad ends in a noticeably pointed nose, and the skin may look somewhat granular and mottled on the underside. The skin forms a fold at the back of the head. The body color is variable both within populations and in individuals, ranging from different shades of light or dark gray and brown to brownish yellow or reddish. A broad, wavy central band sometimes runs from the nose or eye region down the entire body, with the outer bands being a different shade. Narrowmouth toads have no tympanum and presumably hear by receiving vibrations to the inner ear through another pathway.

WHAT DO THE TADPOLES LOOK LIKE? The tadpoles may be an inch long and are always less than 2 inches. Most are black (sometimes dark brown) above and below, often with a white stripe extending partway down the middle on each side of the tail. Viewed from above, the body shape is somewhat like a square with rounded corners; viewed from the side, the body is noticeably flattened compared with tadpoles of other species.

Tadpoles often have a white stripe extending partway down the middle on each side of the tail.

How do you identify an eastern narrowmouth toad?

SKIN
Smooth

LEGS
Short

FEET AND TOES
Toes not webbed

BODY PATTERN AND COLOR
Gray to brown, often with a broad, dark band along middle of back

DISTINCTIVE CHARACTERS
Fat, rounded body; fold of skin across head behind the eyes; distinctly pointed nose; no tympanum

CALL
Drawn-out, raspy, nasal *baa* or *meh*

SIZE
max tadpole = 2"
typical adult = 1"

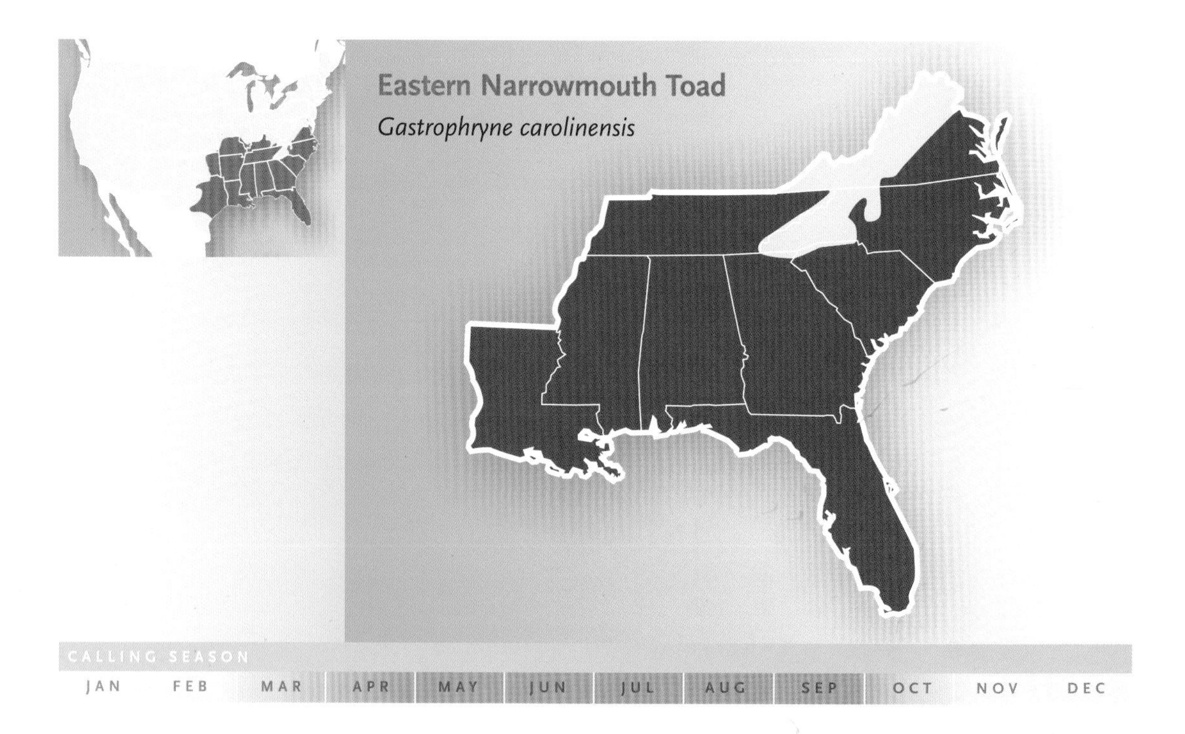

The fold of skin behind the head is characteristic of the eastern narrowmouth toad.

SIMILAR SPECIES Eastern narrowmouth toads are among the most easily distinguished species of eastern frogs and toads. They cannot be mistaken for adults of any of the larger frogs or toads because their maximum size is less than 2 inches. Likewise, the overall shape—round body and very pointed nose—resembles a triangle attached to a circle. The fold of skin behind the head and lack of toe webbing further separate this species from most others within its eastern range.

DISTRIBUTION AND HABITAT The geographic range, which includes part or all of every southern state, extends from Maryland to southern Missouri, exclusive of the Appalachian Mountains. It includes eastern Oklahoma and Virginia and all states to the east. Narrowmouth toads are found in a wide variety of soil types, including sand and organic soils, and forest types, including both pine and hardwood. They occur in moderately urbanized areas, but in all situations they are most likely to be found in moist habitats such as underground burrows or beneath leaf litter, logs, rocks, or other ground cover. Individuals can often be found hundreds of yards from their breeding site.

BEHAVIOR AND ACTIVITY Eastern narrowmouth toads are rarely seen above the ground during daytime except while calling during warm rains. Individuals may be seen at night actively foraging. They can be extremely

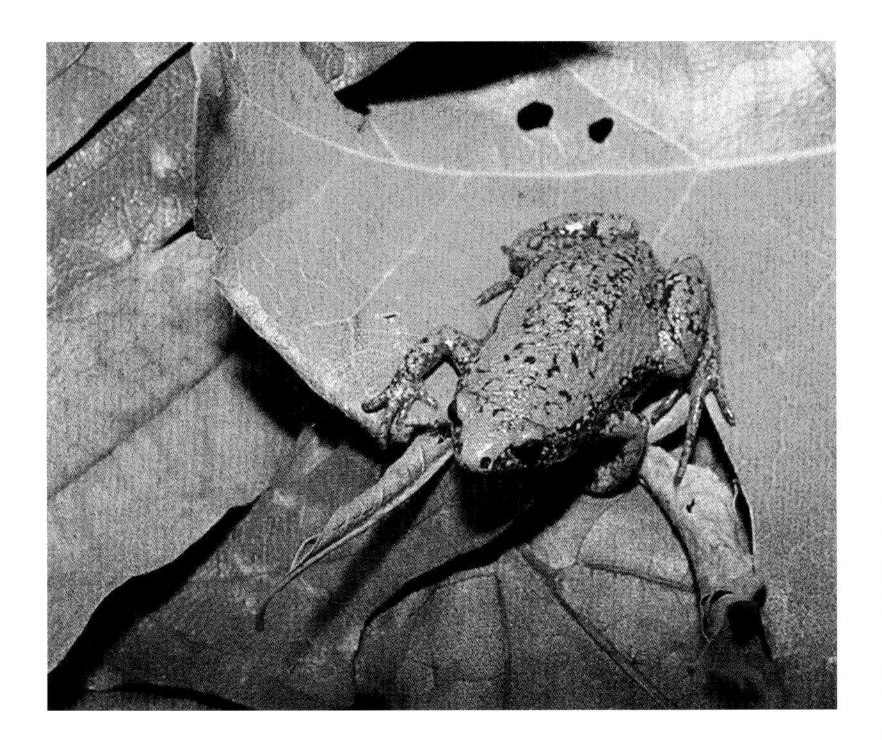

The eastern narrowmouth toad's body color ranges from different shades of light or dark gray and brown to brownish yellow or reddish.

An eastern narrowmouth toad calling

abundant during breeding periods following rains. When disturbed during the day, escaping individuals sometimes look as if they are rapidly walking or scampering away rather than hopping in the style of other frogs and toads.

FOOD AND FEEDING Adults eat various types of small insects but are especially noted for eating ants. Whether or not narrowmouth toads eat imported fire ants is of particular interest because of the widespread environmental impacts of this invasive insect. Some researchers have suggested that the toad can extend the fold of skin on the head forward to keep ants out of its eyes while eating, with the fold in essence functioning as a large eyelid. The tiny tadpoles lack the toothlike mouthparts found in other southeastern tadpoles and eat primarily plankton suspended in the water column.

DESCRIPTION OF CALL The whining *baaaa* of these little toads during breeding was best described by Joseph LeConte, who in the mid-1800s said that the call "exactly resembles the bleating of a lamb." When starting their call, eastern narrowmouth toads often utter a quick whistle that can be heard only at close range.

REPRODUCTION Narrowmouth toads typically breed during warm, rainy weather from late spring to early fall, although males may begin calling ear-

Eastern narrowmouth toads in amplexus

lier in the spring and may continue later in the fall. Males call mostly at night from clumps of grass or other vegetation but occasionally call in the daytime during rainy spells. The eggs are laid in temporary aquatic sites that include roadside ditches, floodplain depressions, and puddles in fields or forests. A female may lay more than 1,000 eggs in batches of approximately 10–100; they float in a single layer like a thin mat on the water's surface. The eggs can hatch within a few days, and the tadpoles develop over a period of about 3–10 weeks, depending on water temperatures. The young tail-less toads are less than half an inch long when they emerge from the water.

The eggs of eastern narrowmouth toads are laid on the water's surface in small groups.

PREDATORS AND DEFENSE Predators of tadpoles presumably include carnivorous aquatic insect larvae and adults, small fish, and small aquatic snakes. The adults are eaten by cattle egrets and several species of snakes, including southern banded watersnakes, glossy crayfish snakes, garter snakes, cottonmouths, and copperheads. Surprisingly, no reports exist of either eastern or southern hognose snakes eating narrowmouth toads, although both snake species specialize on toads (*Bufo* and *Scaphiopus*). This suggests that narrowmouth toads may be unpalatable to hognose snakes, which sometimes share the toads' habitats. The slimy skin secretion can cause a burning sensation if it contacts the mucous membranes and is probably toxic to many potential predators. Adults' most effective defense against birds and many other predators is probably their restricted activity during daytime hours.

CONSERVATION The greatest threats to eastern narrowmouth toads are habitat fragmentation and loss. These toads often call from roadside ditches but can suffer road mortality when they must travel across highways with heavy nighttime traffic to reach breeding areas. Urban development, as a whole, also causes habitat fragmentation and results in destruction of the toads' natural habitats.

The greenhouse frog's overall color is typically brownish or reddish.

Greenhouse Frog *Eleutherodactylus planirostris*

DESCRIPTION Greenhouse frogs are very small; even large adults are only slightly over an inch in total length. The overall body color is brownish or reddish, but individuals have one of two distinct appearances: either dark mottling on the back or two light-colored stripes that extend lengthwise down the body. The legs have dark bands, which may be indistinct in some individuals. The eyes are typically red, and the belly is white or grayish white and may have tiny dark spots. The toes of the hind feet completely lack webbing.

WHAT DO THE TADPOLES LOOK LIKE? This species does not have a tadpole stage. See "Reproduction" below.

Greenhouse frogs are most often found close to the ground.

How do you identify a greenhouse frog?

SKIN
Rough

LEGS
Moderately robust

FEET AND TOES
No webbing between toes

BODY PATTERN AND COLOR
Back brownish with dark mottling or light stripes

DISTINCTIVE CHARACTERS
Small size; reddish eyes in many individuals; complete lack of webbing on feet

CALL
Chirp repeated four to six times

SIZE
max tadpole = n/a
typical adult = 1"

The toes of the greenhouse frog's hind feet completely lack webbing.

The legs have dark bands, though these may be indistinct in some individuals.

SIMILAR SPECIES The small size and complete absence of webbing between the toes will eliminate adults of this species from most other frogs in the Southeast. The reddish eyes distinguish it from the Puerto Rican coqui in the area of southern Florida where both occur.

DISTRIBUTION AND HABITAT Greenhouse frogs occur naturally in Cuba, the Bahamas, and possibly other islands in the West Indies, and the species has been known from southern Florida since at least 1875. By the early 2000s greenhouse frogs had dispersed throughout most of Florida, including all of the Keys and parts of the panhandle. They have also been reported from scattered localities in Louisiana, southern Alabama, and coastal and southern Georgia. The species was reported from New Orleans as early as the 1970s, and prior to Hurricane Katrina in 2005 was known from several locations around the city. At the time of this writing, the status of greenhouse frogs and other amphibians in New Orleans is unknown. Individuals occasionally climb, but this frog is most often found close to the ground. Greenhouse frogs are usually found in warm, humid areas and can occupy a wide range of terrestrial habitats if moist ground cover or other suitable hiding places are available. In this regard, they can fare extremely well in vegetated areas or gardens in urban and suburban areas and in natural woodlands.

BEHAVIOR AND ACTIVITY Greenhouse frogs hide in moist retreats in crevices or under leaves, logs, or other ground litter; they will even hide in gopher tortoise burrows. They commonly venture out at night, during rain, or in humid, cloudy weather. They can be very abundant in some locations. Although the typical home range size is unknown, the species is noted for its ability to disperse. Individuals hitchhike from one area to another in potted plants, gardening supplies, and other household items where they are often overlooked because of their small size. Greenhouse frogs are active in all months in southern Florida if temperatures are warm and conditions humid. They are noted for emerging and calling during the day in response to lawn watering.

Some greenhouse frogs are grayish.

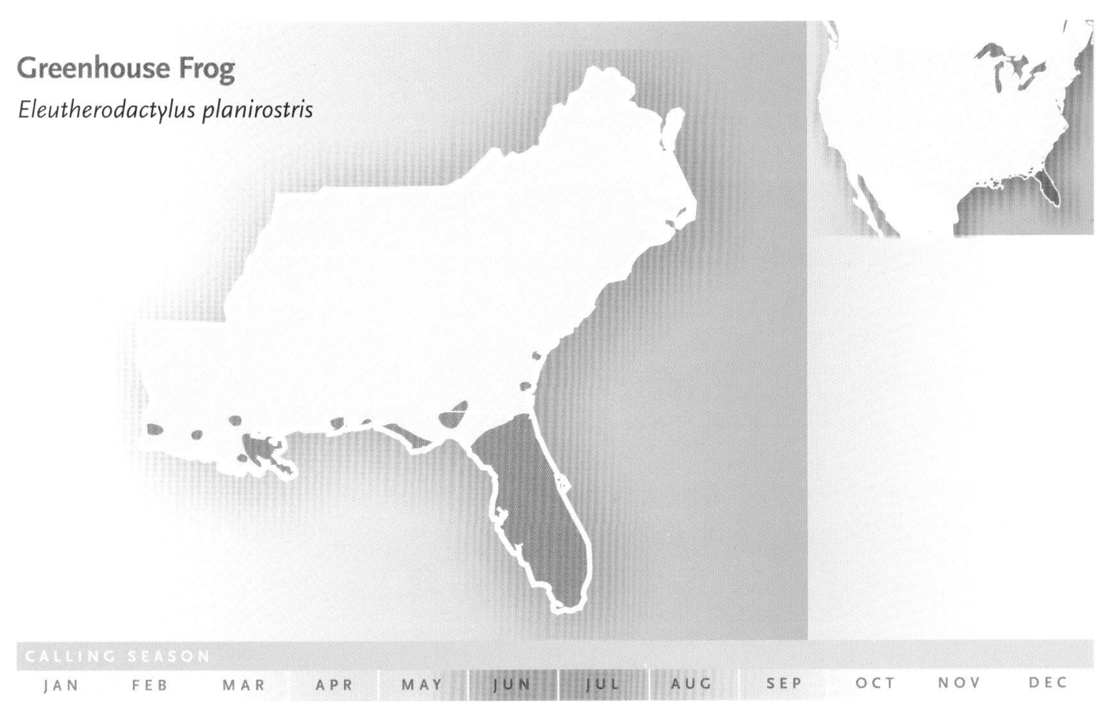

FOOD AND FEEDING Their small size limits greenhouse frogs to small invertebrates such as ants, beetles, spiders, earthworms, and mites. The young froglets presumably eat smaller individuals of these forms of invertebrate prey.

Did you know?

Some frogs lay their eggs on land. In the Southeast, the greenhouse frog and the coqui both lay their eggs in moist places where the eggs develop into tiny, fully formed frogs before hatching.

DESCRIPTION OF CALL The advertisement call is a series of 4–6 soft chirping notes that resemble those of a small bird.

REPRODUCTION Male greenhouse frogs typically call from February to November in southern Florida, but the breeding season is truncated at both ends at higher latitudes; they breed only from April to September in the central part of the state. Males usually call on rainy nights. Females lay their eggs on land, usually in a humid spot and often under debris. In Florida, the average clutch size is about 16 eggs, the upper limit being 26. The larvae develop directly inside the eggs—within less than 2 weeks if conditions are warm and humid—and the young emerge with a nub of a tail and a total length of less than a quarter of an inch.

PREDATORS AND DEFENSE Known predators include Cuban treefrogs and ringneck snakes, but other species of frog-eating snakes, frogs, toads, and birds are also likely to prey on greenhouse frogs. Their primary forms of defense are being secretive during the day and jumping to escape potential predators.

Greenhouse frogs usually have a mottled appearance. Some individuals have two light-colored stripes that extend lengthwise down the body.

CONSERVATION As is true for most invasive species, the primary conservation concern with the greenhouse frog is its effects on native species in the parts of Florida where large populations have become established. Although no substantial negative impacts have been documented, greenhouse frogs may compete with other insect-eating species for food resources, potentially harming populations of some native amphibians or small reptiles. The ease with which greenhouse frogs can disperse as a result of human activities increases colonization opportunities in places that are often far from other colonies.

COMMENTS In 2006, amphibian biologists placed this species in the genus *Euhyas.*

Coquis are small frogs that have been introduced into a limited number of places in Florida.

Puerto Rican Coqui *Eleutherodactylus coqui*

DESCRIPTION The Puerto Rican coqui is a small frog. In Florida, adult females, which grow larger than males, are only about 1.75 inches long; in their native Puerto Rico adults can be more than 2 inches long. The basic body color is highly variable but is some shade of dark or light brown or gray. Darker spots, irregular stripes, or other markings may be visible on some individuals; others are a solid color. The belly is lighter than the back, ranging from white to yellowish with dark stippling. The eyes are brown or yellowish brown, but not reddish. The toes of the hind feet are not webbed. Relatively large, flat toe pads are present.

WHAT DO THE TADPOLES LOOK LIKE? This species does not have a tadpole stage. See "Reproduction" below.

SIMILAR SPECIES The complete absence of webbing between the toes will distinguish adults of this species from most other frogs of similar size in the Southeast. Greenhouse frogs in the area of southern Florida where both species occur usually have reddish eyes.

DISTRIBUTION AND HABITAT The natural geographic range of the coqui is confined to Puerto Rico, but in the mid-1970s the species was known to be present at a single location in Dade County, Florida. This original

How do you identify a Puerto Rican coqui?

SKIN
Rough

LEGS
Moderately robust hind legs

FEET AND TOES
No webbing between toes; toe pads present

BODY PATTERN AND COLOR
Dark brown, tan, or gray, with or without darker markings

DISTINCTIVE CHARACTERS
Small size; brown to yellowish brown eyes; complete lack of webbing on feet

CALL
High-pitched two-note chirp

SIZE
max tadpole = n/a
typical adult = 2"

population was probably eliminated a few years later by a hard freeze in the Miami area. Coquis appeared again in extreme southern Florida in the early 2000s, closely associated with greenhouses, especially those with bromeliads. They were also reported to have been introduced into Louisiana, in the New Orleans area, but their status had not been confirmed a year after the flooding of the city following Hurricane Katrina in 2005.

BEHAVIOR AND ACTIVITY Coquis characteristically remain hidden during the day under ground litter such as dead leaves, rocks, or wood; some hide above the ground in crevices, treeholes, or the leaves of plants. In Florida, they typically do not venture far from greenhouses, which have the warm, moist conditions these little frogs require. They are most active at night or, sometimes, on cloudy days. They do not migrate, and individuals typically remain within an area of only a few square feet, although they will climb small trees, shrubs, and vines several feet above the ground. Both sexes are territorial and defend their daytime retreats and feeding areas from other coquis—and sometimes even lizards or humans—with warning calls. A defending coqui may use physical aggression, such as wrestling or biting, if another coqui ignores the warning call.

The coqui typically has a light band between the eyes.

FOOD AND FEEDING Coquis eat almost any small invertebrates they can catch, including insects and spiders, and are also known to eat the eggs of other coquis.

DESCRIPTION OF CALL The advertisement call consists of two loud, very high-pitched chirps. The common name derives from the distinctive *co* and *qui* sounds. Warning calls are more variable, often given in more rapid succession than the advertisement call, and with the *qui* note occurring more frequently than the *co*.

Coquis have been introduced into Hawaii as well as southern Florida.

Puerto Rican Coqui

Eleutherodactylus coqui

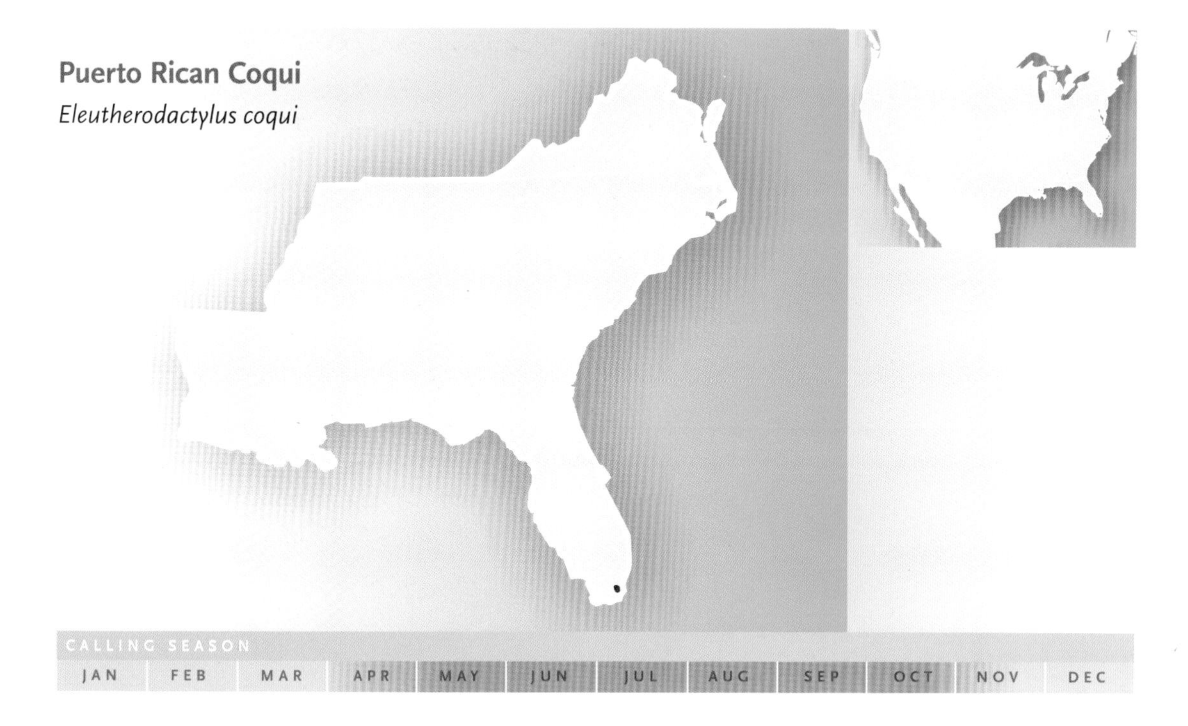

REPRODUCTION Puerto Rican coquis probably breed in every month in Florida, with the peak season being May–October. Males call from trees or other vegetation a few inches to several feet above the ground. When a female responds, the male leads her to the terrestrial site where she will deposit her eggs. Afterward, the male parent stays with and guards the one to two dozen eggs (the average clutch size is higher in Puerto Rico). The eggs develop into larvae that undergo metamorphosis within the egg and emerge from it within several days as quarter-inch-long froglets. The newborn froglets have a short tail that soon disappears.

PREDATORS AND DEFENSE A variety of large invertebrates such as spiders and scorpions prey on coquis, as will lizards, snakes, and birds if given the opportunity. Cuban treefrogs are considered likely predators in Florida because they eat other frogs and are often associated with greenhouses.

CONSERVATION The Puerto Rican coqui has not become widespread or abundant enough anywhere in the Southeast to warrant concern that it is having serious negative impacts on native amphibians. Coquis are viewed as a nuisance in Hawaii, primarily because of their vocalizations, which some people find annoying. This has not been considered a major problem in Florida, perhaps due to the species' limited distribution.

A researcher removes frogs from a bucket trap along a drift fence. Drift fences allow researchers to study amphibians as they move into and out of wetlands.

Research museums are extremely valuable resources for herpetologists.

People and Frogs and Toads

WHAT IS A HERPETOLOGIST?

A herpetologist is a scientist who studies amphibians and reptiles, which collectively are called herpetofauna. Like other scientists who work with particular groups of organisms, most herpetologists specialize in a particular field of biological study. Herpetologists at universities, museums, and research institutes attempt to answer questions about the origins of particular families or species and their genetic relationships, morphology, physiology, and patterns of behavior. Ecology is a common area of herpetological research. Research conducted in these broad areas is often applicable to other groups of animals as well.

Since the late 1900s, many herpetologists have focused on conservation issues. Amphibians in particular have been studied extensively to examine the impacts of land management practices—especially forestry programs—on animal populations. Herpetologists have also used amphibians as model species to examine the effects of pesticides on organisms because they are particularly sensitive to these chemicals. Countless other research projects involve amphibians and reptiles. In fact, several national and international scientific journals are devoted entirely to publishing herpetological research.

Barking treefrog floating in a wetland

A researcher swabbing a frog to detect chytrid fungus, a disease that has resulted in the decline or extinction of many species.

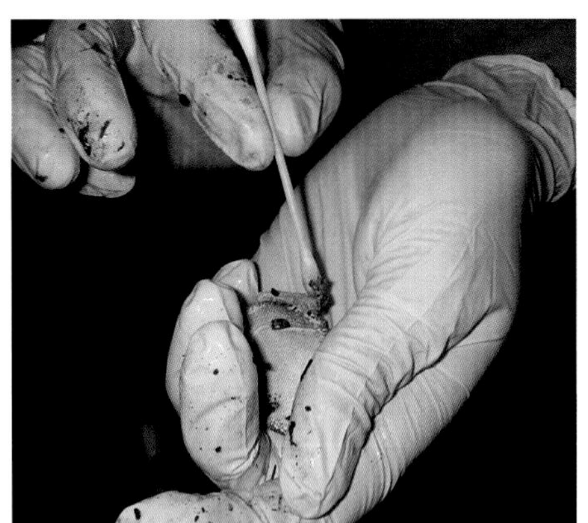

Why Do Herpetologists Study Frogs and Toads?

Herpetologists study frogs and toads in the Southeast and other parts of the world for a variety of reasons. Many ecological studies of frogs and toads focus on issues related to the global decline of amphibians and conservation concerns. Some frogs and toads are sensitive indicators of environmental quality and are considered to be sentinels of ecological hazards that may not be readily detectable by other means. Studies of the basic ecology and behavior of species recognized as threatened or endangered are often necessary to understand the causes of declines and to design measures to preserve the species and their habitats.

Anurans have special traits, such as vocalization and absorption of air and water through the skin, that allow behaviorists to address questions related to animal acoustics and physiologists to pursue questions about animal respiration and water regulation. Medical researchers have used the skin secretions of poison dart frogs to develop anesthetics more powerful than morphine. High school students usually get their first lesson in anatomy and physiology by dissecting a frog.

Did you know?

Because of a fungal disease, frog populations in the highlands of Central America have declined precipitously.

What Techniques Do Herpetologists Use to Study Frogs and Toads?

Researchers have developed special techniques to observe, capture, and study many groups of animals. Frogs and toads are used in many different kinds of laboratory experiments, in part because some species can be easily kept in captivity. Vocal animals like frogs and toads are also ideal for studies in sound labs with special auditory equipment. Tadpoles have been the subjects of many laboratory experiments because they are easy and inexpensive to maintain in aquaria. Many species of frogs and toads can be bred in captivity as well as being easily captured in the field.

Among the most effective ways to capture large numbers of frogs and toads for laboratory experiments or ecological field studies is to use a drift fence. The fence itself is simply a "wall" of aluminum sheeting or silt fencing that is erected in an area where frogs and toads occur. Funnel traps or buckets are placed at intervals along the fence. Amphibians that encounter the drift fence tend to turn and follow it, and eventually enter a funnel trap or fall into a bucket. In studies focusing on frogs and toads, drift fences are commonly placed between an aquatic breeding site and the upland habitat of the species of interest. Drift fence studies in the Southeast have yielded more than a dozen frog and toad species and 20,000 individuals

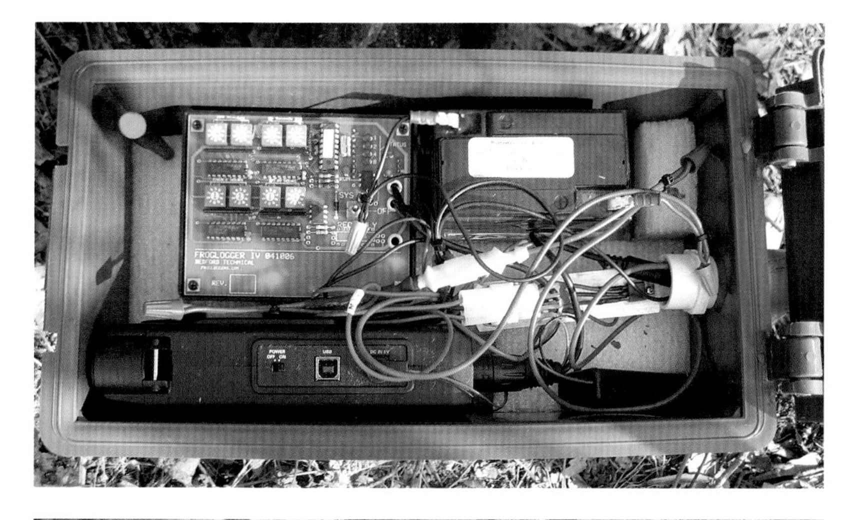

Automated recording systems, sometimes called "frogloggers," record sound automatically and can assist researchers in detecting rare species of frogs and toads.

on a single rainy night! Tadpoles for use in laboratory or field studies are easy to capture with dip nets or minnow traps set in their wetland.

Because most frogs and toads are very vocal creatures, several research techniques take advantage of their calls to locate individuals or choruses at night. Calling surveys (e.g., North American Amphibian Monitoring Program) in which investigators record times, locations, and weather conditions when certain species are breeding can provide valuable ecological information on the distribution and abundance of both rare and common species. The development of the "froglogger," an automated recording device that can be left for long periods in remote locations, has allowed researchers to determine calling patterns without being constantly present.

Some researchers use PVC pipes to study treefrog ecology in wetlands. Many treefrogs seek refuge inside the pipes.

A drift fence works as a barrier. When frogs and toads run into the fence, they are directed toward traps that are checked on a daily basis.

A Cope's gray treefrog leaves its refuge inside a PVC pipe.

Because the eggs of different species are distinctive, systematic surveys of egg masses at breeding sites can be a particularly effective way to determine the extent of successful breeding during a given season. And because southeastern anurans typically lay large numbers of eggs, researchers can remove eggs for experiments without jeopardizing local populations. Likewise, the numbers, kinds, and stages of development of tadpoles can reveal a great deal about the ecology of a wetland and the species of frogs and toads that inhabit it.

Adult frogs captured in a field study can be measured and marked for future identification. Information gained from their later recapture allows researchers to determine growth rates and distances moved between captures and to estimate population sizes. Frogs and toads can be marked for individual identification in several ways. The most common marking technique is to clip unique combinations of toes—taking care, of course, not to remove critical toes. For example, males of many southeastern frog species have enlarged thumbs for breeding purposes. A less invasive marking technique is the use of colored elastomers. The elastomer compound is injected in liquid form under the skin and then hardens and remains as a visible marker. Different combinations of colors, locations, and configurations can be used to give unique marks to numerous individual frogs or toads. Some biologists inject tiny fluorescent tags with numbers on them under the skin of the frog. The tags can be read through the translucent skin with the aid of an ultraviolet light and a magnifying glass. Amphibian biologists have also used passive integrated transponders, or PIT tags, microchips that are injected into the animal's body cavity and then later read by a device similar to a grocery store's bar code reader.

Did you know?

Scientists have described more than 5,000 species of frogs and toads, and more are discovered every year.

Researchers can use fluorescent powder of different colors to track the movements of frogs and toads several hundred feet. This toad is covered in green fluorescent powder that will fall off as it moves from place to place, allowing its trail to be followed at night with the use of a UV (ultraviolet) flashlight.

Researchers inject tiny numeric tags under the skin of frogs to identify study animals.

To follow the movement of frogs and toads over distances up to a mile or more, researchers attach radiotransmitters around their waist.

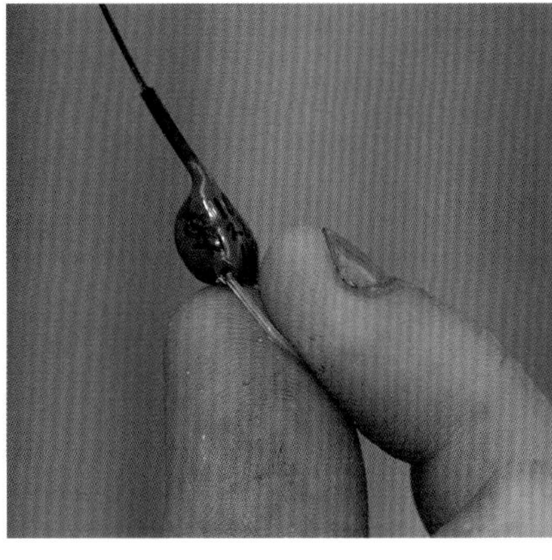

A radiotransmitter used to track the movements of frogs

Each PIT tag has a number that individually identifies a particular frog or toad.

Radiotelemetry can be used to study the movements and behavior of free-ranging frogs and toads in the field. A small radio transmitter attached to the animal's back emits a signal at a particular radio frequency that allows a researcher with a receiver and a special antenna to follow it and study its activity, movement patterns, and habitat use. Amphibians too small to carry a transmitter can be dusted with fluorescent powder. The powder falls off as the animal moves, leaving a trail that can be followed using an ultraviolet light.

What Rules Must Herpetologists Follow to Study Frogs and Toads?

Herpetologists who collect animals for research or educational purposes, such as for a funded research project or a college class in herpetology, must be aware of the state and federal laws that protect wildlife in their region. All states in the Southeast require researchers to have a scientific collecting permit if they plan to remove frogs and toads from the wild for laboratory experiments or to mark and measure individuals in the field. The dusky gopher frog, which is protected under the U.S. Endangered Species Act, can be handled in the field or kept in captivity *only* by someone with a special federal permit. Wildlife laws differ from state to state, and learning what they are for a given area can be time-consuming. Obtaining permits may require paperwork and even payment of a fee, but most rules and regulations are intended to protect the animals.

Wetlands that sometimes dry out provide critical habitat for many species of frogs and toads.

CONSERVATION

Aside from their fascinating looks, sounds, and intriguing biological traits, why are frogs and toads important? Why should we support conservation programs that protect their wetland breeding habitats and the uplands where many of them live the rest of the year? Why should we be concerned that some scientists are predicting the extinction of one-third of the world's frogs and toads within the next few decades?

In addition to contributing an appealing component to the biodiversity of the Southeast, frogs and toads are bioindicators or biomonitors of environmental integrity; that is, their absence or decline in abundance may signal general problems in the environment. For example, many frogs and toads cannot survive in degraded wetlands that have been contaminated by pesticides. These unhealthy habitats often feature high numbers of diseased or abnormal frogs (or no frogs at all) that can indicate underlying problems in the environment that are not otherwise obvious. Additionally, because adult frogs and toads are predators, and thus higher on the food chain than plant-eating animals, toxins tend to accumulate in their tissues and are more easily detected than they are in organisms lower on the food chain. Finally, because frogs and toads have a two-phased life cycle, living both in the water and on land, they are exposed to a wider range of environmental stressors.

Roadside ditches can provide important fish-free habitat in which frogs and toads can reproduce.

This bullfrog tadpole has deformities that could have been caused by pollution.

In sum, anurans as a group are vital elements of the Southeast's natural ecosystems. During a single breeding season, frogs and toads in a southeastern wetland can lay more than a million eggs that result in literally tons of offspring. Such productivity ultimately feeds hundreds of species of invertebrate and vertebrate predators, with aquatic predators eating eggs and tadpoles, and terrestrial ones eating the young and adults. Nonetheless, in a healthy ecosystem, their populations are often abundant even after the predators have taken their toll.

Adult frogs and toads are also effective predators that help keep other animal populations in check. In only one year, the frogs and toads of a single wetland can consume millions of insects. Some of the larger species such as bullfrogs, pig frogs, and some of the true toads extend their diet to include small fish, snakes, birds, and mammals.

What Are the Causes of Global Declines of Amphibians?

Local populations and entire species of frogs and toads are declining worldwide faster than any animal group has declined since the dinosaurs disap-

Did you know?

Malformations in which frogs have extra legs, missing legs, or other deformities occur in unusually high numbers in some populations. Scientists have determined that pollution, parasites, and nutritional deficiencies may be causes of these malformations.

peared. The geographic reach of this problem includes the United States, and the seriousness could become compounded in the Southeast, which has a far greater diversity of amphibian species than any other region of the country. In addition to the unquestionable decline of some species, scientists have documented strange body malformations including extra limbs, missing limbs, and other deformities. The exact causes of the malformations and declines have often been difficult to determine, but they almost invariably have a common denominator: human activities.

Habitat destruction is by far the primary conservation concern for southeastern frogs and toads. Many areas of the Southeast are developing at an alarming rate. Although a few anuran species may persist in urban and suburban areas, most die out when their wetland and/or upland habitats are altered significantly or destroyed. Roads, parking lots, and commercial structures separate the remaining pieces of habitat, often isolating populations of terrestrial species from their wetland breeding areas. Mortality on roads is a continual threat to adult frogs and toads in all parts of the Southeast as they travel to breeding sites during rainy periods.

Fire suppression has resulted in major changes to ecosystems that are important for many species of frogs and toads. Longleaf pine habitats in particular require fire to control undergrowth. Ornate chorus frogs, which inhabit wetlands within longleaf pine ecosystems, have declined substantially in many areas, and some herpetologists believe fire suppression may be the reason.

Another major cause of decline in the numbers and health of amphibian populations is pollution of aquatic habitats by toxic chemicals. Frogs and

Pollution can cause deformities, such as the missing eye of this bullfrog.

Automobiles kill many species of frogs and toads (left). Habitat destruction (right) is the primary conservation concern throughout much of the Southeast.

Preservation of wetlands (left) is critical for the persistence of many frog and toad populations. Prescribed burning (right) is an essential management tool in much of the Southeast.

Amphibians inhabiting wetlands near industrial facilities may be affected by pollution. This barking treefrog lives in a wetland near a coal power plant and may suffer from effects of heavy metals generated by the plant.

toads respire and absorb moisture through their skin, but during these processes they also absorb chemicals, and that makes them particularly vulnerable to toxins in water, air, and soil. Some chemicals mimic the hormone estrogen and feminize male frogs, making them unable to reproduce. The sources of toxic chemicals are nearly endless. Some originate from pesticides, herbicides, and fertilizers that enter the soil and water as urban or agricultural run-off. Others come from industrial sources that result in acid rain or toxic airborne chemicals. Many of these deadly products end up in wetlands and can kill eggs, tadpoles, or adults outright or harm populations by interfering with reproduction, limb development, and the ability to resist disease.

During the 1990s, amphibian biologists found that parasites and pathogens were having greater impacts on frogs and toads than ever before. Pollution and environmental contaminants that compromised the health of individuals were at least partly responsible. The deadly fungus chytridiomycosis, or chytrid for short, infects the skin of frogs and toads. Chytrid is apparently responsible for the extinction of many species of frogs and toads worldwide and has had especially devastating impacts in Australia and the highlands of Central America, the latter an area with one of the most diverse frog and toad faunas in the world. Following an outbreak of chytrid at study sites once teeming with diverse assemblages of frogs and toads, researchers have returned to find only dead and dying frogs or no frogs at all. Whether human activities contribute to these fungal outbreaks remains to be determined.

Other threats to our frogs and toads include the introduction of species such as predatory fish, which eat the eggs and tadpoles of many species, and human-subsidized predators such as house cats that are allowed to roam free and prey on native species. The nonnative Cuban treefrog and cane toad may outcompete native frogs and toads for food and habitat, or may prey directly on eggs, tadpoles, or adults. Ultraviolet radiation resulting from ozone depletion has the potential to harm some species by increasing egg mortality. Finally, the effects of global climate change are already being felt and may eventually have devastating impacts on many species of frogs and toads.

Cuban treefrogs are apparently expanding their range and may pose problems for some native wildlife.

Although fun to catch, introduced fish, such as this largemouth bass, can wreak havoc on native amphibian populations.

One of the few wetlands where dusky gopher frogs are still known to breed

A dusky gopher frog, a critically endangered species

What Can You Do about Frog and Toad Conservation?

Persuading legislators to pass laws that protect nongame species is difficult, but it can be done; the Endangered Species Act of 1973 is an example. Unfortunately, the primary threats to the frogs and toads of the Southeast are difficult to regulate. The main culprits appear to be loss or degradation of suitable habitat as a consequence of commercial development or the release of toxic chemicals into the environment. Although scientists have confirmed the loss or decline of several species, only a single species of southeastern frog, the dusky gopher frog, is on the federal endangered species list. For example, by the early 2000s, the river frog had disappeared from its former range in North Carolina and the common gopher frog had been documented from only a single site in South Carolina for most of the previous decade. Yet the federal government has failed to recognize the plight of these species by offering significant protection.

As things stand, state and local legislators are unlikely to enact laws or regulations to prevent habitat from being destroyed solely because a particular wetland or upland area is an important environment for frogs or toads. But anurans, sensitive barometers of environmental conditions, should perhaps become our gauge for what level of environmental insult a habitat can withstand. If the frogs and toads stop calling, it is only a matter of time before people are affected as well.

Conservation Education

The first step in developing an effective conservation approach for any wildlife group is to educate the public about the threats faced by the species of interest. Toward this end, educate yourself and then help inform others of the steps required to keep the frogs and toads of the Southeast a healthy and viable part of our natural heritage. Become involved with national conservation groups that focus on amphibian conservation, such as Partners in Amphibian and Reptile Conservation (PARC); participate in frog and toad monitoring projects such as the North American Amphibian Monitoring Program (NAAMP); or join a local or regional herpetology group with interests in conservation. Participating in wetland restoration projects or other activities that preserve wetlands and their surrounding terrestrial habitat can help greatly in the conservation of frogs, toads, and a variety of other wildlife species. Finally, continue to learn as much as you can about the frogs and toads that inhabit your area. Look for them. Listen for them. Becoming aware of their presence will increase your appreciation for them and inspire you to encourage others to do what they can to keep anurans an integral part of our natural environments.

Members of Partners in Amphibian and Reptile Conservation (PARC) work toward a common goal of preserving amphibians and reptiles.

FROGS AND TOADS AS PETS

Anyone who acquires a frog or toad as a pet should know how to care for it in captivity, make sure no laws are violated when obtaining it, make sure the animal is healthy and free of noticeable disease, and understand that frogs and toads kept as pets should not be released into the wild. Some species of frogs and toads make very good pets; others do not. Some have a rather fragile or delicate nature, some have very specific habitat requirements, and some are just unable to adapt to captive conditions. Other species might survive if properly cared for, but are so secretive that they might never be seen. In general, frogs and toads are not personable pets that can be handled frequently like hamsters or parakeets.

Selecting a Pet Frog or Toad

Species that do make good pets include several of the larger North American treefrogs, such as gray treefrogs, barking treefrogs, and green treefrogs. Many treefrogs are brightly colored and boldly patterned, and they usually do fine in a terrarium with potted plants and a bowl of clean water. Other species that generally do well in captivity include several of the toads, such as Fowler's toads and American toads, although many people consider them less attractive and thus less desirable as pets. Some true frogs, such as bullfrogs and leopard frogs, are commonly kept as pets, but these species often injure themselves by jumping into the walls of their cage. Several species of frogs not native to the Southeast, including White's treefrogs (*Litoria caerulea*), pixie frogs (genus *Pyxicephalus*), and horned frogs (genus *Cera-*

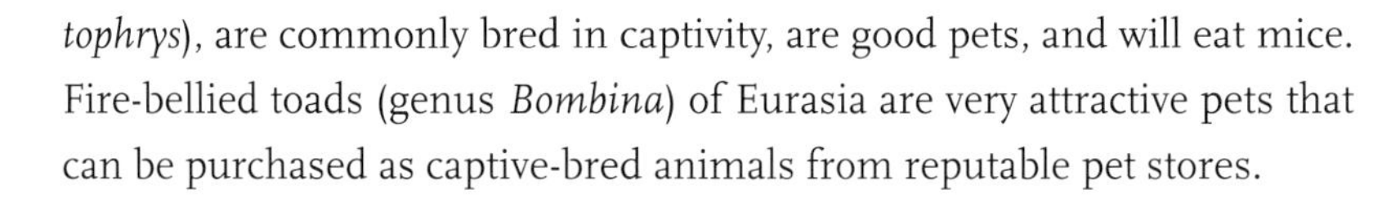

tophrys), are commonly bred in captivity, are good pets, and will eat mice. Fire-bellied toads (genus *Bombina*) of Eurasia are very attractive pets that can be purchased as captive-bred animals from reputable pet stores.

Acquiring a Pet Frog or Toad

Frogs and toads can be captured in their natural habitats or acquired from a pet store or breeder. In general, captive-bred frogs make better pets than wild-caught individuals. Before purchasing a frog or toad from a pet store, ask whether it is wild-caught or captive-bred. Animals in pet stores are not always well cared for, so inspect the animal carefully to be certain that it is healthy. Many captive-bred frogs and toads can be purchased directly from breeders. Such animals are usually much healthier than those from pet stores, and are usually less expensive as well.

Capturing a frog or toad is often more fun than actually keeping it as a pet. Treefrogs and toads can be collected in many areas during their breeding season. In general, removing one or a few individuals has a negligible impact on the overall population, but before you collect any frog or toad, you should be aware of and abide by any federal, state, or local laws that might restrict collection. It is also crucial to understand that by removing a frog or toad from the wild and keeping it in captivity, you assume an ethical responsibility to properly care for that animal in the best way possible. If you are unable or unwilling to do so, it is wise to consider another pet option.

Albino frogs and toads, such as this albino bullfrog (below top), are often kept as pets because they are unusual and many people consider them attractive.

Fire-bellied toads (below middle) from Eurasia are popular pets.

White's treefrogs (below bottom) sometimes get large enough to eat small mice.

Caring for Frogs and Toads in Captivity

Most captive frogs and toads can be fed insects, such as crickets purchased from pet stores. Many herp enthusiasts recommend "dusting" crickets with calcium and vitamins to ensure that the pet is getting a complete complement of required nutrients. Some larger frogs will eat mice and other small vertebrates. Keeping the cages of frogs and toads as clean as possible is vital not only for the health of the animals but also for the health of the owner. A dirty environment encourages the growth of salmonella, bacteria that can cause serious illness in humans. Maintaining the proper temperature is important for most frogs and toads; many species require higher humidity than is typically found within a home and must be misted daily. Regardless of what species you decide to keep, it is your responsibility to learn about that species and its specific needs. Many books and

A terrarium set up to house treefrogs

Web sites provide excellent information for anyone interested in keeping pet frogs and toads.

Releasing Pet Frogs and Toads

Some frogs and toads can live a surprisingly long time. Individuals of some species can live more than 15 years in captivity. Too often, though, pet owners tire of having a pet and want to rid themselves of the responsibility for its care. Returning wild-caught animals to the wild may be possible, but you must first consider several factors. Can you return it to the same place you captured it? In general, wild-caught animals should be released only at the exact point of capture. Even if you can release the animal where it was captured, it could have contracted a disease in captivity that could affect the entire population of the species in that area. For that reason, many scientists recommend against releasing any captive frog or toad. Captive-bred or exotic species should never be released in the wild.

Potential Concerns

Captive frogs and toads, like many other animals, can pose dangers to their owners. Some frogs and toads can develop diseases in captivity that may be transmissible to humans. Always wash your hands carefully with soap before and after handling any frog or toad. Many frogs and toads have skin toxins that can cause irritation to the eyes, nose, and mouth. Thus, if you handle a frog or a toad, you should always do so gently, reducing the stress on the animal and the amount of toxin it releases. Do not touch your eyes, nose, or mouth after handling an amphibian until you wash your hands thoroughly with soap.

Calling Months

The time of year or day and the weather conditions affect calling activity. In general, relatively warm, wet nights are the best time to hear frog choruses, though many winter-calling species call during the daytime. Some species of frogs and toads may call at any time of the year if the weather conditions are right.

COMMON NAME	JAN	FEB	MAR	APR	MAY	JUN	JUL	AUG	SEP	OCT	NOV	DEC
CRICKET FROGS, CHORUS FROGS, AND TREEFROGS												
Northern Cricket Frog												
Southern Cricket Frog												
Upland Chorus Frog												
Southern Chorus Frog												
Brimley's Chorus Frog												
Little Grass Frog												
Ornate Chorus Frog												
Strecker's Chorus Frog												
Mountain Chorus Frog												
Spring Peeper												
Green Treefrog												
Pine Barrens Treefrog												
Barking Treefrog												
Common Gray Treefrog												
Cope's Gray Treefrog												
Bird-voiced Treefrog												
Squirrel Treefrog												
Pine Woods Treefrog												
Cuban Treefrog												
TRUE FROGS (RANIDAE)												
Southern Leopard Frog												
Pickerel Frog												
Green or Bronze Frog												
Wood Frog												
Carpenter Frog												
Florida Bog Frog												
Bullfrog												
Pig Frog												
River Frog												
Crawfish Frog												
Gopher Frog												
Dusky Gopher Frog												
TRUE TOADS (BUFONIDAE)												
American Toad												
Fowler's Toad												
Southern Toad												
Coastal Plain Toad												
Oak Toad												
Cane Toad												
SPADEFOOT TOADS (SCAPHIOPIDAE)												
Eastern Spadefoot												
Hurter's Spadefoot												
NARROWMOUTH TOADS (MICROHYLIDAE)												
Eastern Narrowmouth Toad												
NEW WORLD TROPICAL FROGS (LEPTODACTYLIDAE)												
Greenhouse Frog												
Puerto Rican Coqui												

Calling can occur anytime during wet, warm weather.

What kinds of frogs and toads are found in your state?

COMMON NAME	VA	NC	SC	GA	FL	TN	AL	MS	LA
CRICKET FROGS, CHORUS FROGS, AND TREEFROGS									
Northern Cricket Frog	●	●	●	●	●	●	●	●	●
Southern Cricket Frog	●	●	●	●	●	●	●	●	●
Upland Chorus Frog	●	●	●	●	●	●	●	●	●
Southern Chorus Frog		●	●	●	●		●	●	
Brimley's Chorus Frog	●	●	●	●					
Little Grass Frog	●	●	●	●	●		●		
Ornate Chorus Frog		●	●	●	●		●	●	●
Strecker's Chorus Frog									●
Mountain Chorus Frog	●	●		●		●	●	●	
Spring Peeper	●	●	●	●	●	●	●	●	●
Green Treefrog	●	●	●	●	●	●	●	●	●
Pine Barrens Treefrog		●	●		●		●		
Barking Treefrog	●	●	●	●	●	●	●	●	●
Common Gray Treefrog	●	●				●			●
Cope's Gray Treefrog	●	●	●	●	●	●	●	●	●
Bird-voiced Treefrog			●	●	●	●	●	●	●
Squirrel Treefrog	●	●	●	●	●		●	●	●
Pine Woods Treefrog	●	●	●	●	●		●	●	●
Cuban Treefrog*					●				
TRUE FROGS (RANIDAE)									
Southern Leopard Frog	●	●	●	●	●	●	●	●	●
Pickerel Frog	●	●	●	●	●	●	●	●	●
Green or Bronze Frog	●	●	●	●	●	●	●	●	●
Wood Frog	●	●	●	●		●	●		
Carpenter Frog	●	●	●	●	●				
Florida Bog Frog					●				
Bullfrog	●	●	●	●	●	●	●	●	●
Pig Frog			●	●	●		●	●	●
River Frog		●	●	●	●		●	●	
Crawfish Frog						●		●	●
Gopher Frog		●	●	●	●	●	●		
Dusky Gopher Frog							●	●	
TRUE TOADS (BUFONIDAE)									
American Toad	●	●	●	●		●	●	●	●
Fowler's Toad	●	●	●	●	●	●	●	●	●
Southern Toad	●	●	●	●	●		●	●	●
Coastal Plain Toad								●	●
Oak Toad	●	●	●	●	●		●	●	●
Cane Toad*					●				
SPADEFOOT TOADS (SCAPHIOPIDAE)									
Eastern Spadefoot	●	●	●	●	●	●	●	●	●
Hurter's Spadefoot									●
NARROWMOUTH TOADS (MICROHYLIDAE)									
Eastern Narrowmouth Toad	●	●	●	●	●	●	●	●	●
NEW WORLD TROPICAL FROGS (LEPTODACTYLIDAE)									
Greenhouse Frog*				●	●		●		●
Puerto Rican Coqui*					●				
TOTAL	25	30	30	31	32	21	31	28	28

* *introduced species*

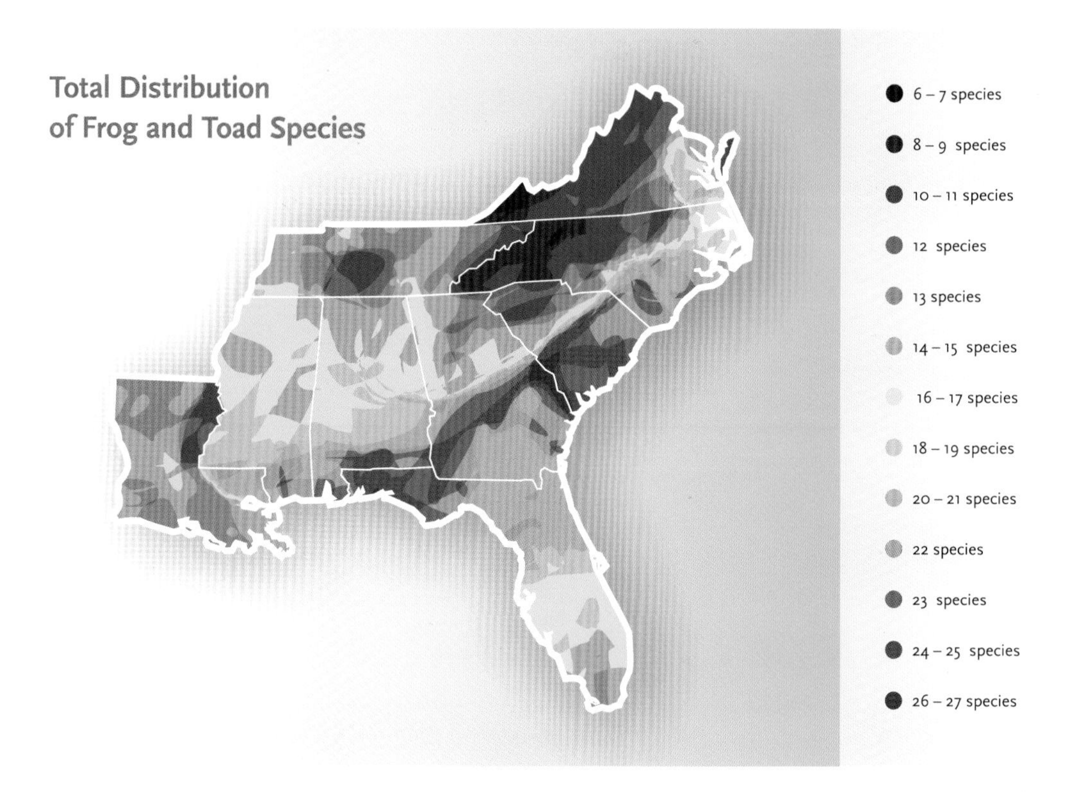
Total Distribution
of Frog and Toad Species
6 – 7 species
8 – 9 species
10 – 11 species
12 species
13 species
14 – 15 species
16 – 17 species
18 – 19 species
20 – 21 species
22 species
23 species
24 – 25 species
26 – 27 species

Glossary

Advertisement call The vocalization given by male frogs during the breeding season to let females know of their availability and location.

Albino An animal completely lacking the pigment that provides the color to skin and eyes. Animals lacking only dark pigment, or melanin, are often referred to as albino but are more properly said to be amelanistic.

Amplexus The grasping of a female frog around the waist or under the arms from behind by a male during mating.

Anterior Referring to the end of the animal toward the head.

Anurans Amphibians that are frogs or toads.

Biodiversity The numbers, distribution, and abundance of species within a given area.

Bioindicator A species whose health or condition, either at the individual level or at the population level, indicates the condition of the habitat or ecosystem as a whole.

Biomass The weight of living things in the environment.

Biomonitor To measure and record the number, types, and characteristics of organisms.

Brummation *See* Hibernation.

Bufotoxin A toxic substance created in the parotoid and skin glands of true toads that is secreted as a defense against predators.

Chytrid A group of soil fungi, one of which has been shown to infect many species of amphibians and has resulted in declines of populations in many parts of the world.

Chytridiomycosis The disease caused by the chytrid fungus that is sometimes fatal to frogs and toads.

Cloaca (*adj.* cloacal) A single opening through which the urinary, digestive, and reproductive tracts exit the body.

Clutch A group of eggs laid together at one time by a single individual.

Cold-blooded A nontechnical term that refers to animals whose body temperature is determined largely by environmental conditions and the thermoregulatory behavior of the animal. *See also* Ectotherm.

Cranial crests Paired solid ridges on top of the head of true toads; typically these are located alongside and behind the eyes and in front of the parotoid glands.

Dorsolateral Referring to the area of the body between the back and sides.

Ecology The study of how organisms interact with their environment.

Ectotherm An animal whose body temperature is largely determined by environmental conditions and the thermoregulatory behavior of the animal. *See also* Cold-blooded.

Elastomers Material injected under the skin of amphibians as a colored liquid that hardens and becomes a visible marker that can be used to identify individuals.

Endangered Referring to a species or population that is considered at risk of becoming extinct.

Endotherm An animal that maintains a constant body temperature primarily through the use of heat generated by its own metabolism. *See also* Warm-blooded.

Ephemeral wetlands Typically small, isolated depressions that hold water in some but not all years or seasons.

Extinct Referring to species with no living individuals.

Fall Line The line separating the Piedmont from the Coastal Plain in the southeastern United States.

Family (of frogs or toads) A taxonomic group containing two or more closely related genera.

Generalist An animal that does not specialize on any particular type of prey or is not restricted to a particular habitat.

Genus (*pl.* genera) A taxonomic grouping of one or more closely related species.

Herpetofauna The amphibians and reptiles that inhabit a given area.

Herpetologist A scientist who studies amphibians and reptiles.

Herpetology The study of amphibians and reptiles.

Hibernation A period of inactivity during cold periods. Also known as "brummation" when referring to amphibians and reptiles.

Hybrid The offspring of mating between two different species.

Intergrade An intermediate form of a species resulting from mating and genetic mixing between individuals of two or more subspecies along a zone where they are contiguous. Intergrade specimens may possess traits of all subspecies involved.

Metamorphosis The process by which a tadpole changes into a frog or toad.

Naiad A dragonfly larva; some prey on tadpoles.

NAAMP North American Amphibian Monitoring Program; a collaborative effort of the U.S. Geological Survey, state natural resources agencies, and nonprofit organizations to monitor populations of calling frogs and toads.

Nocturnal Active at night.

Orthopteran A member of the insect order Orthoptera, which includes the katydids, grasshoppers, and crickets.

Oxbow A U-shaped lake that was once the main channel of a nearby river.

PARC Partners in Amphibian and Reptile Conservation; the largest partnership group dedicated to the conservation of all amphibians and reptiles and their habitats.

Parotoid glands A pair of external glands located behind the head on toads that secrete bufotoxin, a poisonous substance, for defensive purposes.

Phylogeny The evolutionary relationships among different groups and species of animals.

PIT tag Acronym for passive integrated transponder tag, a glass-encapsulated electronic device injected into the body cavity of animals for identification purposes. PIT tags emit signals that can be read by a special device in close proximity.

Posterior Away from the head of an animal and toward the tail.

Radiotelemetry A tracking method using a radiotransmitter attached to or implanted in an animal and a directional antenna and radio receiver.

Release call A short call given by a male grasped accidentally by another male during breeding or by a female not interested in mating.

Satellite male A male frog or toad that stays in the vicinity of a calling male and tries to intercept and mate with females that are attracted to the calling male.

Silviculture The development, care, and management of forests for timber production.

Specialist An animal restricted in its choice of diet or habitat.

Species Typically an identifiable and distinct group of organisms whose members breed and produce viable offspring under natural conditions.

Subspecies A taxonomic unit, or "race," within a species, usually defined as morphologically distinct and occupying a geographic range that does not overlap with that of other races of the species. Subspecies may interbreed naturally in areas of geographic contact. *See also* Intergrade.

Taxonomy The scientific field of classification and naming of organisms.

Tubercles Small, smooth wartlike structures on the upper body of some toads.

Tympanum (*pl.* tympanums, tympana) The externally visible eardrum of frogs and toads.

Warm-blooded A nontechnical term that refers to an animal that maintains its body temperature primarily through the use of metabolic heat. *See also* Endotherm.

Further Reading

Bartlett, R. D., and P. P. Bartlett. 1998. *A Field Guide to Florida Reptiles and Amphibians.* Houston: Gulf Publishing Company.

Beltz, E. 2005. *Frogs: Inside Their Remarkable World.* Buffalo, N.Y.: Firefly Books.

Conant, R., and J. T. Collins. 1991. *A Field Guide to Reptiles and Amphibians of Eastern and Central North America.* 3rd ed. Boston: Houghton Mifflin.

Dodd, C. K., Jr. 2004. *The Amphibians of Great Smoky Mountains National Park.* Knoxville: University of Tennessee Press.

Dorcas, M. E., S. J. Price, J. C Beane, and S. S. Cross. 2007. *The Frogs and Toads of North Carolina.* North Carolina Wildlife Resources Commission, Raleigh, N.C.

Duellman, W. E., and L. Trueb. 1994. *Biology of Amphibians.* Baltimore: Johns Hopkins University Press.

Dundee, H. A., and D. A. Rossman. 1996. *Amphibians and Reptiles of Louisiana.* Baton Rouge: Louisiana State University Press.

Gibbons, J. W., and R. D. Semlitsch. 1991. *Guide to the Reptiles and Amphibians of the Savannah River Site.* Athens: University of Georgia Press.

Jensen, J. B., C. D. Camp, J. W. Gibbons, and M. J. Elliott, eds. 2008. *Amphibians and Reptiles of Georgia.* Athens: University of Georgia Press.

Lannoo, M., ed. 2005. *Amphibian Declines: The Conservation Status of United States Species.* Berkeley: University of California Press.

Martof, B. S., W. M. Palmer, J. R. Bailey, J. R. Harrison III, and J. Dermid. 1980. *Amphibians and Reptiles of the Carolinas and Virginia.* Chapel Hill: University of North Carolina Press.

Meshaka, W. E., Jr., B. P. Butterfield, and J. B. Hauge. 2005. *Exotic Amphibians and Reptiles of Florida.* Melbourne, Fla.: Krieger Publishing Company.

Mount, R. M. 1975. *The Reptiles and Amphibians of Alabama.* Tuscaloosa: University of Alabama Press.

Semlitsch, R. D., ed. 2003. *Amphibian Conservation.* Washington, D.C.: Smithsonian Institution Press.

Stebbins, R. C., and N. W. Cohen. 1995. *A Natural History of Amphibians.* Princeton: Princeton University Press.

Acknowledgments

We appreciate the support and understanding of our families during the preparation of the book: Tammy, Taylor, Jessika, and Zachary Dorcas; Carolyn, Laura, Michael, Jennifer, Allison, and Parker Gibbons; Susan Lane and Keith Harris; and Jennifer and Jim High.

We appreciate the constructive comments on the species accounts from the following people: Steve Bennett, Kristen Cecala, George Cline, Gabrielle Graeter, Leigh Anne Harden, Trip Lamb, Bruce Means, Joe Mendelson, Walter Meshaka, Tony Mills, Paul Moler, Emily Moriarty, Priya Nanjappa, John Palis, Tom Pauley, Steven Price, Fred Punzo, Stephen Richter, David Scott, and Brian Todd. Steven Price, Tony Mills, and Charlotte Steelman assisted us by reviewing numerous other parts of the book. Craig Guyer and Joseph Mitchell provided thorough reviews of the entire manuscript and Pete Wyatt gave advice on the conservation listing of *Rana capito* in Tennessee. John Jensen supplied distribution maps and other information for species found in Georgia. Ronn Altig, Steve Bennett, Carlos Camp, Jessika Dorcas, Julian Harrison, Bob Jones, Steve Price, Brian Sullivan, Jeff Boundy, Ken Marion, Mark Bailey, Floyd Scott, Paul Moler, Joe Mitchell, Nathan Parker, and Evan Eskew all assisted by reviewing the maps. Kevin Samples generated the species density map. Jason Norman, Tony Mills, Trip Lamb, and Steven Price provided technical advice on particular species. Evan Eskew reviewed the calling season charts. Margaret Wead, Teresa Carroll, and Michelle Kirlin scanned and organized images for the book. Michelle Kirlin assisted with the production of the maps and helped organize distribution data. Steven Price reviewed the first set of page proofs, although we take responsibility for any remaining errors.

We are very grateful to the many herpetologists throughout the country who provided us with a spectacular array of anuran slides and images for use in the book. The excellent color images generously offered by so many experts in animal photography have added tremendously to the value of this book. We thank the following for providing images and slides for the book: Richard Bartlett, Steve Bennett, Jonathan Campbell, Kristen Cecala, Evan Eskew, Gabrielle Graeter, Kristine Grayson, Aubrey Heupel, E. Pierson Hill, Trip Lamb, Tom Luhring, Chris McEwen, Tom McNamara, Carl Mehling, Shannon Pittman, Bob Rothman, Zbynek Rocek of Charles University in Prague, David Scott, Michael Sisson, Brian Todd, R. Wayne Van Devender, J. D. Willson, and Robert Zappalorti.

Support and resources for MED while writing this book were provided by the Department of Biology at Davidson College and the National Science Foundation.

Finally, because the book supports the efforts of Partners in Amphibian and Reptile Conservation to promote education about reptiles and amphibians, we thank the many PARC members who offered encouragement, advice, and enthusiastic support.

Credits

The authors would like to thank the following individuals and organizations for providing photographs:

Chris Austin
Photograph on page 24.

Randy Babb
Photograph on page 130 (bottom).

Richard D. Bartlett
Photographs on pages 30, 43, 55, 65 (both), 66 (top), 68, 74, 76, 104 (top), 111, 115 (top), 129, 136 (bottom), 139 (both), 146, 151 (both), 169, 170, 175, 177, 178, 186, 187, 188, 189 (top), 193 (top), 197, 198 (both), 214 (left), and 216 (bottom).

Steve Bennett
Photographs on pages vi, 23, 106, 133, 142 (top), 179, and 190.

Kurt A. Buhlmann
Photograph on page 148 (top).

Kristen Cecala
Photographs on pages 40 (top) and 77.

Mark Danaher
Photograph on page 100.

Michael E. Dorcas
Photographs on pages 1 (bottom), 17, 22 (both), 117 (bottom), 201 (right), 202, 204 (bottom), and 211 (bottom right).

Sarah Durant
Photograph on page 213 (bottom).

Charlie Eichelberger
Photograph on page 12 (third on left).

Evan A. Eskew
Photographs on pages 164, 176, and 208.

Whit Gibbons
Photograph on page 201 (left).

Gabrielle Graeter
Photographs on pages 206 (top) and 207 (left).

Kristine Grayson
Photographs on pages 12 (bottom right) and 124 (top).

Jeff Hall
Photographs on pages 115 (bottom) and 131.

Aubrey M. Heupel
Photographs on pages 11 (bottom), 16 (bottom), 19 (top), 26, 27 (top), 37, 39 (top), 49, 50 (top), 54, 70, 79, 81 (both), 82, 83 (top), 92, 93, 97, 98, 99 (bottom), 109 (top), 137, and 191 (left).

E. Pierson Hill
Photographs on pages 127 and 183.

Marcus Jones / istockphoto.com
Photograph on page ii (top).

Trip Lamb
Photographs on pages 10 (top), 14, 15 (bottom), 64 (top), 73 (left), 78, and 163.

Thomas M. Luhring
Photographs on pages 4 (top), 8, 9 (top), 11 (top left and right), 12 (top two), 15 (top), 16 (top), 25 (bottom and top left), 27 (bottom), 28 (both), 30 (top), 32, 33, 45 (bottom), 58 (top), 59, 75 (top), 84 (bottom), 87 (bottom), 91, 94 (right), 101, 126, 138, 147, 150, 168, 191 (right), 192 (both), 200, 207 (right), 209 (top), 211 (left top and bottom), 212, 216 (middle), and 217.

John MacKay
Photograph on page 10 (bottom).

Barry Mansell
Photographs on pages 130 (top) and 152 (right).

Chris McEwen
Photographs on pages 71 (bottom) and 155.

Tom McNamara
Photograph on page 4 (bottom).

Tony Mills
Photograph on page 9 (bottom left).

Jeffrey Peter / istockphoto.com
Photograph on page viii.

Shannon Pittman
Photographs on pages 205 and 206 (bottom).

Roger Rittmaster
Photograph on page 58 (bottom).

Zbyněk Roček
Photograph on page 2.

David E. Scott
Photograph on page 9 (bottom right).

Michael Sisson
Photographs on pages 152 (left), 153, 154, and 214 (right).

Dieter Spears / istockphoto.com
Photograph on pages ii–iii.

Charlotte Steelman

Photographs on pages 203 (both) and 204 (top).

R. Wayne Van Devender

Photographs on pages 12 (bottom left), 13 (first and third), 19 (bottom), 20, 25 (top right), 40 (bottom), 41 (bottom), 44, 48, 52, 53, 57 (bottom), 60, 63 (right), 64 (bottom), 66 (bottom), 69, 71 (top), 72, 73 (right), 83 (bottom), 86, 87 (top), 89, 95, 96 (both), 99 (top), 103, 105, 109 (bottom), 113 (both), 116, 118 (bottom), 122 (top), 128 (both), 136 (top), 141, 143, 145 (both), 148 (bottom), 158 (top), 160, 161 (top), 165 (top), 171, 172 (top), 181 (bottom), 185 (bottom), 189 (bottom), 194 (top), 195, 196, and 209 (bottom).

John White

Photograph on page 210.

John D. Willson

Photographs on pages i, 3 (all), 5, 6 (top left and right), 7, 13 (second and fourth), 18 (both), 21, 29, 39 (bottom), 42, 45 (top), 46, 47, 50 (bottom), 51 (both), 56, 57 (top), 61 (both), 62, 63 (left), 102, 110 (both), 112, 114, 118 (top), 120, 121 (both), 122 (bottom), 125, 132, 134, 135, 142 (bottom), 157 (both), 158 (bottom), 161 (bottom), 162, 166, 167, 181 (top), 182 (both), 184, 185 (top), and 211 (top right).

Robert Zappalorti

Photographs on pages vii, 1 (top), 6 (bottom), 75 (bottom), 84 (top), 90 (both), 94 (left), 104 (bottom), 117 (top), 124 (bottom), 140, 165 (bottom), 172 (bottom), 174, 193 (bottom), 194 (bottom), 213 (top), and 216 (top).

Index of Scientific Names

Boldface page numbers refer to species accounts.

Index of Common Names

Boldface page numbers refer to species accounts. *Italicized* page numbers refer to illustrations.